*BY THE SAME AUTHOR.*

---

## LIMES, HYDRAULIC CEMENTS AND MORTARS.

CONTAINING

NUMEROUS EXPERIMENTS CONDUCTED IN NEW YORK CITY.

Fifth Edition, with illustrations, 8vo, cloth, $4.00.

---

OFFICIAL REPORT OF OPERATIONS

AGAINST THE

## DEFENCES OF CHARLESTON HARBOR, 1863.

COMPRISING

THE DESCENT UPON MORRIS ISLAND, THE DEMOLITION OF
FORT SUMTER, AND THE SIEGE AND REDUCTION
OF FORTS WAGNER AND GREGG, WITH
70 LITHOGRAPHIC PLATES,
VIEWS, MAPS, ETC.

8vo, cloth, $10.00. Half Russia, $12.00.

SUPPLEMENTARY REPORT TO THE ABOVE.

8vo, cloth, $5.00. Half Russia, $7.00.

---

COIGNET BETON, AND OTHER ARTIFICIAL STONE.

Illustrated by 9 Plates, Views, etc., 8vo, cloth, $2.50.

---

## ON STRENGTH OF THE BUILDING STONES

IN THE UNITED STATES.

8vo, cloth, $1.50.

# A PRACTICAL TREATISE

ON

# ROADS, STREETS, AND PAVEMENTS.

BY

Q. A. GILLMORE, A.M.,

LIEUT. COL. U. S. CORPS OF ENGINEERS, BREVET MAJOR-GENERAL U. S. ARMY.

AUTHOR OF "LIMES, HYDRAULIC CEMENTS, AND MORTARS," "BÉTON AGGLOMÉRÉ AND OTHER ARTIFICIAL STONES," "REPORT ON STRENGTH OF BUILDING STONES OF THE UNITED STATES," &c., &c.

NEW YORK

D. VAN NOSTRAND, PUBLISHER,

23 MURRAY AND 27 WARREN STREETS.

1876.

NEWBURGH STEREOTYPE COMPANY.

# PREFATORY NOTE.

In the preparation of the following pages a few leading objects have been kept in view by the author.

*First.* To give, within the compass of one small volume, such descriptions of the various methods of locating country roads, and of constructing the road and street coverings in more or less common use at the present day, as will render the essential details of those methods, as well as certain improvements thereon of which many of them are believed to be susceptible, familiar to any intelligent non-professional reader.

*Second.* To make such practical suggestions with respect to the selection and application of materials, more especially those, with the properties and uses of which builders are presumed to be the least acquainted, as seem needful in order to develop their greatest practical worth, and realize their greatest endurance.

*Third.* To institute a just and discriminating comparison of the respective merits of the several street pavements now competing for popular recognition and favor, under the varying conditions of traffic, climate, and locality, to which they are commonly subjected.

Q. A. G.

NEW YORK, March 1, 1876.

# TABLE OF CONTENTS.

## CHAPTER I.

### LOCATION AND GRADES OF COUNTRY ROADS.



## CHAPTER II.

### EARTHWORK, DRAINAGE AND TRANSVERSE FORM OF COUNTRY ROADS.

## CHAPTER III.

### ROAD COVERINGS.



## CHAPTER IV.

### MAINTENANCE AND REPAIRS OF ROADS.

## CHAPTER V.

### STREETS AND STREET PAVEMENTS.

## CHAPTER VI.

### SIDEWALKS AND FOOTPATHS.



## CHAPTER VII.

### TRAMWAYS AND STREET RAILWAYS.

PRACTICAL TREATISE

ON

# ROADS, STREETS, AND PAVEMENTS.

## CHAPTER I.

### LOCATION AND GRADES OF COUNTRY ROADS.

### Considerations Governing Location.

THE considerations which should govern the Engineer in locating the line of an ordinary wagon road are (1) the present and prospective amount of traffic over the road; (2) its general character, whether light or heavy; (3) the convenience and necessities of the community tributary to the line; and (4) the natural features of the country through which the road must pass. The labor of the preliminary examination of the ground will be considerably lessened by keeping in view a few elementary principles, viz: (1) that the natural water courses are not only the lowest lines, but the lines of the greatest longitudinal slope in the valleys through which they flow; (2) that the direction and position of the principal streams give also the direction and approximate position of the high ground or ridges which lie between them; and (3) that the positions of the tributaries to the larger streams generally indicate the points of greatest depression in the summits of the ridges, and therefore the points at which lateral communication across the high ground separating contiguous valleys could be most readily made.

## Reconnaissance.

With the aid of an ordinary map of the country, if reasonably correct, it is entirely practicable to trace upon it with a sufficient degree of accuracy for the immediate object in view, not only the general directions of the ridges or highest ground, but also to locate approximately those routes most suitable for the purposes of a road across the hills, from one valley to another.

Being provided with the information usually supplied by maps or, in the absence of trustworthy maps, having secured that information by an instrumental examination, the topographical and other characteristic features of the ground should be carefully studied by travelling in both directions over the several routes, upon any one of which the line may be located, carefully noting down for future comparison, the distinctive features of each.

## Aneroid Barometer.

If the line passes over such hilly or undulating ground, that considerable differences of level are necessarily encountered in its location, valuable aid may be derived from a pocket Aneroid barometer. This instrument shown in section through the axis in Fig. 1, consists of a flat cylindrical box A, exhausted of air, the top of which is thin metal corrugated in concentric circles so as to render it quite elastic. As the atmospheric pressure increases, the elastic top of the box is forced in or down, and as it decreases it is forced out or up. This movement of the top of the box due to changes in the atmospheric pressure, is conveyed by multiplying levers DE, EG, GH, and a small chain II to an index needle NN, moving over a circular scale MM, graduated to correspond with the standard mercurial barometer. The spiral spring S, by its tension raises the long arm of the lever DE, when the pressure on the top of the box is lessened, thus keeping the short arm of the lever constantly in contact with the fulcrum C. The Aneroid is used by the following rule: The sum of the readings at two stations, is to their difference, as 55,000 (or twice the height of the atmosphere in feet), is to the ele-

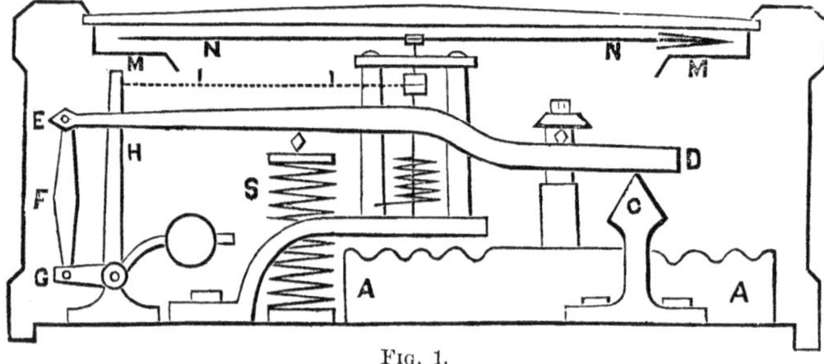


FIG. 1.

vation required. Thus, if the reading at the foot of a hill is 30.05, and at the top 29.44, we have the following 59.49 : 0.61 : : 55,000ft. : 564ft.

By the intelligent use of this barometer, the scope of inquiry may frequently be much narrowed at the outset, and the labor and expense of the subsequent survey greatly abridged. For instance, if the line of communication is to connect two valleys by crossing the high ground between them, it should be located, other things being equal, in the lowest depression of the summit. A reconnaissance with a barometer should indicate with an error not exceeding 10 to 15ft. the relative altitude of the several summit depressions, and therefore the best location of the route, so far as the question of grade fixes the location. The average of several observations with the barometer, is desirable.

## Surveys, Map and Descriptive Memoir.

The reconnaissance having been completed, and the inquiry narrowed down by the exclusion of the least practicable routes, preliminary surveys should then be made of the several *trial-lines,* with a view to determine their length, direction, and position, together with a longitudinal section and numerous cross-sections of each line. All this should be done with sufficient accuracy and minuteness of detail to form the basis of comparative estimates of their practicability and cost. Money liberally spent in surveys entrusted to skillful persons, is wisely spent, and offers the surest guarantee against subsequent mistakes and errors of judgment. Its amount, at the outside, cannot exceed a very small percentage of the cost of constructing the road, while the results, judiciously employed, are sure to furnish innumerable suggestions for lessening the cost without impairing the excellence of the communication.

## Map.

The data obtained from the surveys should be carefully embodied in a map, showing with considerable detail, the topography of the country embraced by the several *trial-lines,* the exact position of these lines, and all the longitudinal and cross-sections. The horizontal scale of the sections should be the same as that of the map, while the vertical scale should be considerably larger, in order to show clearly all the inequalities of the ground. With a horizontal scale of 500ft. to the inch, the vertical scale may be only 20ft. to the inch.

## Descriptive Memoir.

The descriptive memoir should give with minuteness all information, such as the nature of the soil, the character of the several cuttings, whether earth or rock, the kind of rock, etc., etc., that cannot be set forth on the map. The importance of carefully noting in the memoir, and as far as practicable upon the map also, all variations of the character of the cuttings through rock, and especially of maintaining a strict distinction between cuttings in rock and cuttings in earth, will be admitted, when it is remembered that excavations in earth can be made at less than one-fourth the cost of excavations in rock.

## Location of the Line.

"In selecting among the different lines of the survey the one most suitable for a common road, the engineer is less restricted, from the nature of the conveyance used, than in any other kind of communication. The main points to which he should confine his attention are, (1) to connect the points of arrival and departure by the shortest or most direct line ; (2) to avoid all unnecessary ascents and descents, or in other words to keep the ascents and descents within the smallest practicable limits ; (3) to adopt such slopes or gradients for the centre line of the road as the kind of conveyance used may require ; (4) to give the centre line such a position with reference to the natural surface of the ground, and the various obstacles to be overcome, that the cost of labor for excavations and embankments required by the gradients adopted, and also the cost of bridges and other accessories, shall be reduced to the smallest amount." (Prof. Mahan.)

Except in a flat and level country, it will seldom be practicable to adopt, for the axis of the road, the shortest line between the points of arrival and departure. Departures from a straight line are determined by a variety of considerations. In crossing a dividing ridge between two valleys, we seek a depression in the summit, in order to avoid expensive cutting, or the alternative of steep or impracticable gradients; in descending a valley longitudinally, we keep well up on the side of the hill, if necessary, to avoid bridging the ravines and secondary water-courses; if we encounter a swamp or shallow pond, we can frequently, by turning to the right or left, entirely omit the construction of an expensive causeway or other road bed, or substitute for it a short bridge over a narrow water-course, with easy approaches on either side; a stream may cross and then re-cross the direct line, forming an elbow which may sometimes be turned without greatly augmenting the length of the road, thus avoiding the construction and maintenance of two bridges; we may turn aside and even bridge a stream in order to get upon that slope of a valley which will give the best exposure of the road surface to sun and wind; or we may lengthen the road for the purpose of procuring better material for its construction.

### Questions of Expediency to be Considered.

Not infrequently other questions not strictly within the province of the engineer claim attention to such degree that although a straight line between the two termini of the road may be the best, considerations of expediency will very properly prevent its adoption. Intermediate communities and towns contiguous to the line may require accommodation, and whether such accommodation shall be afforded by lateral branch roads, leaving the main line essentially straight, or by running the latter through the several centres of business, will have to be determined upon principles more or less independent of the bare problem of construction and maintenance. For example take the simplest case of three towns A, B, C, Fig. 2, situated upon a uniformly level plain, where the cost of constructing and maintaining a line of communication will be directly proportional to its length. Suppose the points of arrival and departure, A and C, to be 120 miles apart, and that B is equi-distant from A and C, but located 20 miles off the direct line AC. If each town sends out 20 tons of freight per day to be distributed equally (10 tons to each) between the other two towns, a simple calculation will show that with a straight road AC, between the extreme points and a perpendicular branch road B′B to the intermediate town, the daily transportation of this 30 tons of freight, will be equivalent to transporting 1 ton 5600 miles, while if one straight road be built from A to B, and another from B to C, the total carriage will be equivalent to conveying 1 ton only 5060 miles. The difference, (equal to 540 tons carried 1 mile) in favor of the lines AB, BC, over the lines AC and B′B is borne unequally by the three towns in proportion of 335 tons to B, and 102½ each to A and C. In the case stated, it is for the mutual advantage of all the three communities to locate the line from A to B and from B to C. But the conditions would be different if the great bulk of the traffic is between the towns of A and C. For example, if those towns exchange 20 tons per day with each

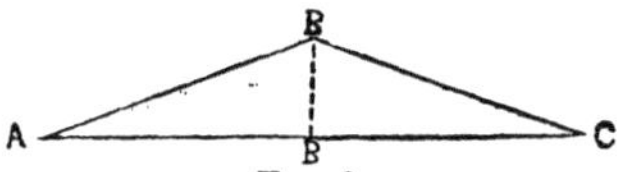

Fig. 2.

other, and only 2 to 3 tons per day with the town B, then the total mileage of transportation necessary in making the interchange, setting other considerations aside, would favor the construction of a straight line AC, between the terminal towns, and a perpendicular branch line B′B to the intermediate town. But as this would require a greater length of road by 13½ miles (140—126½) than a continuous line A B C passing through the intermediate town, the expediency of building and maintaining upwards of 10 per cent more road than is absolutely necessary to connect the three towns, presents itself as a question of some commercial importance, and before adopting the longer system, it should be made quite clear that the yearly saving in cost of transportation will be more than sufficient to pay the interest on the first cost of constructing 13½ miles of road, as well as the annual expense of its maintenance. Except in extreme cases where the two towns at the ends of the line are large in comparison with the intermediate towns, it will usually be found to be most conducive to the convenience of the general public to run the line through all the principal communities, in preference to the plan of communicating with the intermediate places by branch roads, which might necessitate branch lines of wagons and coaches, connecting with those of the main line, attended by all the usual inconvenience and expense of transferring passengers and goods at the points of junction.

In the general case, however, the road will not traverse a level plain but will cross hills, ravines, rivers, and other accidents of the ground, so that the proper solution of the problem will involve a variety of considerations, among which the engineer's ideal of a straight and level line, the wants of the communities to be accommodated, and economy in cost of construction will generally be more or less at variance.

As it should be the first business of the engineer to make himself thoroughly familiar with the character of the country adjacent to the line, for some distance on either side, all the preliminary field work should be directed to that end.

## Contour Lines.

In laying out important roads, and especially in locating streets in towns or thickly settled districts, where the questions of gradient, drainage, sewerage, and water supply assume special importance, it would be well to place the contour lines, or curves of uniform level, upon the map. These curves represent the intersections of the surface of the ground with a series of horizontal planes at equal distances apart, of say 3ft. 5ft. 10ft. or more—and indicate at once to the practiced eye the topography of the country which they embrace. We give, Fig. 3, the contour lines of a small tract of country, showing a variety of natural features, such as undulating slopes, steep hill sides, ravines, water, and marsh. Every curve represents a level line traced upon the ground, the vertical distance, or difference of level, between the curves being 3 feet.

## Parallel Cross Sections.

A survey so complete as would be required for mapping the contour lines in the manner shown in Fig. 3, is seldom, if ever, resorted to, and is indeed unnecessary for the proper location of a country road. It will suffice to take a series of cross sections parallel to each other, and extending a sufficient distance laterally to embrace the width of the country

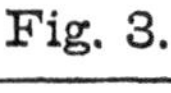
Fig. 3.

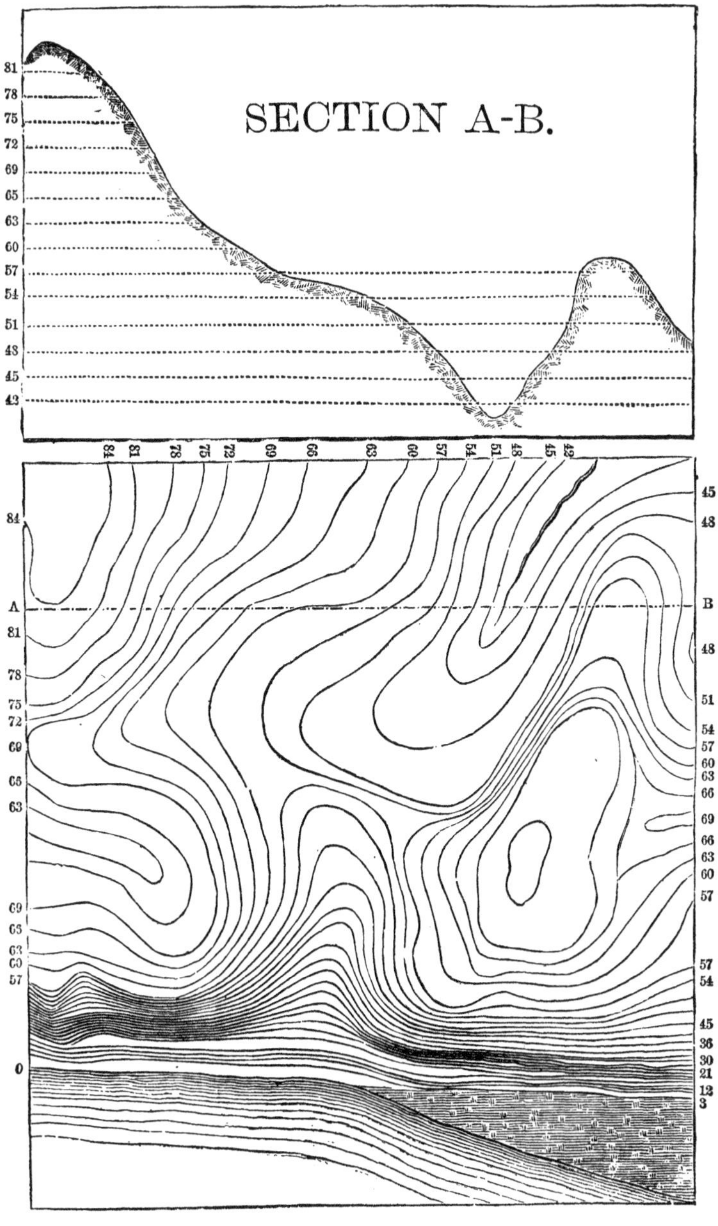

SECTION A-B.

under examination. By plotting these sections to a scale, in their true relative positions, all referred to the same level or datum line, it will be easy to locate the axis of the road properly thereon, and to estimate the quantities of excavation and embankment, provided the sections are taken sufficiently near together. Their distances apart should of course be less in proportion to the ruggedness and unevenness of the country. Suppose, for example, that C, D, M, N, Fig. 4, represents a portion of the strip of country under examina-

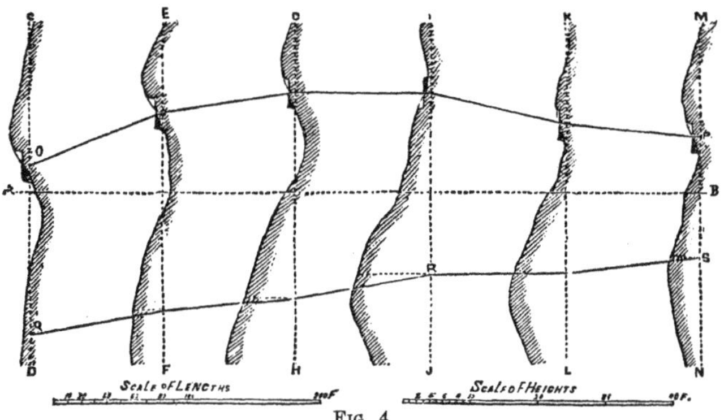


FIG. 4.

tion and A, B, the general direction of the road, which must be located somewhere between the lines C, M, and D, N. Having run the several lines of levels C, D, E, F, etc., transversely, and at least one line A, B, longitudinally so as to establish a common datum line for all, the sections are drawn as shown in the figure. As a part of the map or plan the right lines C, D, E, F, etc., show the positions of the cross sections, while they are also the datum lines, all on the same level, of their respective sections. Those portions of each section to the left of the datum line represent ground above that line and those portions to the right, below it, and a line like O, P, drawn through points in the several sections that are at the same distance from and on the same side of the datum lines, will show the exact location of a level line traced on the surface of the ground. Hence the axis of the road if located on the line O, P, will be level, while upon the line Q, S, it would have an ascending grade of $\frac{1}{30}$ from Q, to R, and a descending grade of $\frac{1}{30}$ from R to S.

Having completed the surveys, and prepared the map and memoir with as much detail as possible, the engineer will then be able to study with intelligence the relative advantages of the trial lines, and to establish definitely their location, direction and gradients, in order that the volumes of excavations and embankments shall balance each other as nearly as possible, except when other methods, hereafter referred to would lessen the expense, while the cost of constructing the necessary bridges, culverts, etc., shall be care fully kept at the minimum.

## Grades.

Upon common roads the grades, or the angles which the axis of the road should make with a horizontal line, depend so much upon the kind of vehicle employed for traffic, the character of road-covering adopted for the surface, and the condition in which that surface is maintained, that no empyrical rule can be laid down. The grade should not be so great as to require the application of brakes to the wheels in descending, or to prevent ordinary vehicles carrying passengers ascending at a trot. In general the gradient should be somewhat less than the *angle of repose*, or that angle upon which the vehicle in a state of rest would not be set in motion by its own weight, but would descend with slow uniform velocity if very slight motion be imparted to it. The grades therefore, suitable for any road, will depend upon the condition, with respect to smoothness and hardness, in which the surface is to be maintained, and hence upon the kind of road-covering used; and as the force of gravity is the same whether the road be rough and soft, or smooth and hard, steep grades are more objectionable upon good roads than upon bad.

## Tractive Force.

Many ingenious experiments have been made at various times to ascertain, in functions of the quality and condition of the road surfaces, the measure of the *tractive force*, or the force required to overcome the resistances which oppose themselves to the movement of a vehicle along horizontal roads of different degrees of smoothness and hardness, and covered with different materials. From some of the experiments of M. Morin, conducted for the French government, the following general results were deduced:

1. The force of traction varies directly with the load and inversely with the diameter of the wheels.

2. The resistance is practically independent of the width of tire on paved or hard Macadamized roads, where that width exceeds 3 or 4 inches.

3. At ordinary walking speed the traction is the same for carriages with springs, as for those without them, the other conditions being the same.

4. The force of traction increases with the speed upon paved, or hard Macadamized roads. When the speed exceeds 2½ miles per hour the increase in the resistance varies directly with the increase in velocity.

Some of M. Morin's results are tabulated below.

| Kind of Road. | Relation of Force of Draught to Weight of Vehicle and Load. | | | | | |
|---|---|---|---|---|---|---|
| | Carts. | Trucks of two tons. | Diligences of five tons. | | Carriages with seats hung on springs. | |
| New road, with gravel covering, 5 inches thick.... | $\frac{1}{12}$ | $\frac{1}{9}$ | $\frac{1}{8}$ | | $\frac{1}{8}$ | |
| Solid earth causeway, with gravel covering 1½ inch thick.................. | $\frac{1}{16}$ | $\frac{1}{11}$ | $\frac{1}{10}$ | | $\frac{1}{10}$ | |
| Earth causeway in very good condition ......... | $\frac{1}{41}$ | $\frac{1}{29}$ | $\frac{1}{26}$ | | $\frac{1}{26}$ | |
| | | | Walk. | Trot. | Walk. | Trot. |
| Broken stone road, very dry and smooth.. | $\frac{1}{75}$ | $\frac{1}{54}$ | $\frac{1}{48}$ | $\frac{1}{41}$ | $\frac{1}{49}$ | $\frac{1}{42}$ |
| Do. moist and dusty... | $\frac{1}{53}$ | $\frac{1}{38}$ | $\frac{1}{34}$ | $\frac{1}{27}$ | $\frac{1}{34}$ | $\frac{1}{27}$ |
| Do. with ruts and mud, | $\frac{1}{33}$ | $\frac{1}{24}$ | $\frac{1}{21}$ | $\frac{1}{18}$ | $\frac{1}{22}$ | $\frac{1}{19}$ |
| Do. with deep ruts and thick mud...... | $\frac{1}{19}$ | $\frac{1}{14}$ | $\frac{1}{12}$ | $\frac{1}{10}$ | $\frac{1}{12}$ | $\frac{1}{10}$ |
| Pavement. Dry ......... | $\frac{1}{90}$ | $\frac{1}{65}$ | $\frac{1}{57}$ | $\frac{1}{38}$ | $\frac{1}{59}$ | $\frac{1}{39}$ |
| Pavement. Muddy....... | $\frac{1}{69}$ | $\frac{1}{50}$ | $\frac{1}{44}$ | $\frac{1}{33}$ | $\frac{1}{45}$ | $\frac{1}{34}$ |

The smoother the road and the less rigid the vehicle, the less will be each equal increase of resistance due to each equal increase of speed.

5. The traction is practically independent of the velocity upon soft dirt and sand roads, or roads freshly and thickly covered with gravel.

6. Upon a smooth-cut, evenly-laid stone pavement, the resistance at a walking speed does not exceed three-fourths that upon the best Macadamized road at the same speed, but at trotting speed it is equal to it.

7. The wear and tear of the road is greater as the diameters of the wheels are less, and is less from vehicles with springs than from those without them.

The following table, resulting from trials made with a dynamometer attached to a wagon moving at a slow pace upon a level, gives the force of traction in pounds upon several kinds of road-surfaces, in a fair condition; the weight of wagon and load being one ton of 2,240 pounds.

| | |
|---|---|
| 1. On best stone trackways | 12½ pounds. |
| 2. A good plank road | 32 to 50 pounds. |
| 3. A cubical block pavement | 32 to 33 " |
| 4. A Macadamized road of small broken stones | 65 pounds. |
| 5. A Telford road, made with six inches of broken stone of great hardness, laid on a foundation of large stones set as a pavement | 46 pounds. |
| 6. A road covered with six inches of broken stone laid on concrete foundation | 46 " |
| 7. A road made with a thick coating of gravel laid on earth | 140 to 147 pounds. |
| 8. A common earth road | 200 pounds. |

In order to apply these results in establishing suitable grades, take the case of the Macadamized road No. 4, in which the tractive force to the gross ton is 65 pounds upon a level road. Let $W=$ the weight of the vehicle and load in pounds; $\mathbf{p}=$ pressure normal to the road-surface in pounds; $\mathbf{t}=$force of traction in pounds on a level road. At the angle of repose of an inclined road, the force, acting parallel to the line of grade, necessary to sustain a carriage and its load in its position on the incline, or to prevent it from moving back by its own weight, is equal to the traction force **t**, which would just move the carriage and load on a level road. Let **h** be the perpendicular and **b** the base of a right angle triangle, of which the hypothenuse B C (Fig. 5) represents the slope of the angle of repose, which somewhat exceeds the greatest admissible gradient. For simplicity, the load may be supposed to rest on a single wheel, shown in the figure. In the smaller similar triangle **t** is the perpendicular,

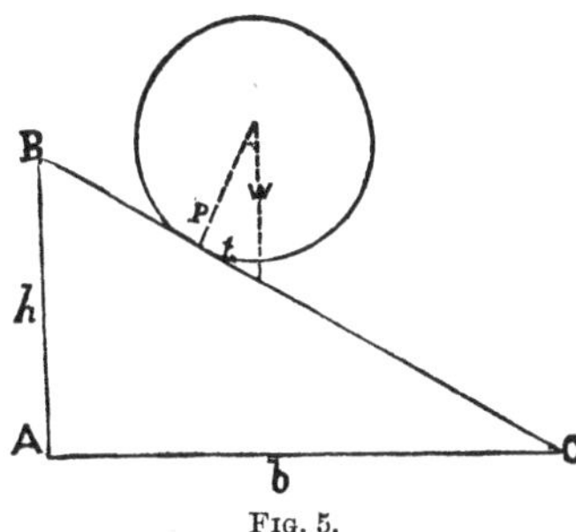

Fig. 5.

**p** the base, and W the hypothenuse, in which $\mathbf{p} = \sqrt{W^2 - \mathbf{t}^2}$.

From the two similar triangles $\mathbf{t} : \mathbf{p} :: \mathbf{h} : \mathbf{b}$, or $\frac{\mathbf{t}}{\mathbf{p}} = \frac{\mathbf{h}}{\mathbf{b}}$, and by substitution $\frac{\mathbf{t}}{\sqrt{W^2 - \mathbf{t}^2}} = \frac{\mathbf{h}}{\mathbf{b}}$. But $\frac{\mathbf{h}}{\mathbf{b}}$, being the perpendicular divided by the base, represents the angle at the base, or the angle of repose, and this is the maximum admissible gradient. Hence the gradient should not exceed the quotient obtained by dividing the force of traction by the square root of the difference between the square of the load and the square of the traction. Upon good roads **t** is so very small in proportion to W that it may be omitted in the denominator, and we have practically for the angle of repose $\frac{\mathbf{t}}{W}$, or the force of traction divided by the weight of vehicle and load.

For road No. 4 the formula becomes $\frac{65}{\sqrt{2240^2 - 65^2}}$, $= \frac{1}{34}$

nearly, indicating that for roads upon which the force of traction per ton is 65 pounds, the grade should be not greater than 1 perpendicular to 34 base; and generally the proper grade for any kind of road, or the ratio of the vertical to the horizontal line, will be equal to the ratio between the force necessary to draw the load and the load itself, upon the same road when level. The grade is usually expressed in the form of a vulgar fraction, having 1 for the numerator, and the horizontal distance corresponding to a rise of one foot for the denominator.

In practice the steepest grades that can be allowed upon Macadamized or Telford roads, in the condition in which their road surface is usually maintained, is about $\frac{1}{20}$, it having been determined by experience that a horse can draw up this slope, unless it be a very long one, his ordinary load for a level road, without the help of a second animal; also that he can attain at a walk, a given height, upon a gradient of $\frac{1}{20}$ without more apparent fatigue, and in nearly the same time that he would require to reach the same height over a proportionately longer road with a slope so gentle—say $\frac{1}{34}$—that he could ascend it at a trot.

It is however more desirable, especially for passenger traffic, to keep the gradients as low as $\frac{1}{34}$, or at the greatest $\frac{1}{30}$, as the maximum slope, whether considered as an *ascent* or a *descent*, so that in the former case the speed need not be slower than a trot, while in the latter it will not be necessary to apply a brake to prevent the load pressing forward upon the horses.

## Undulating Grades.

It is claimed, as having been demonstrated by experience, that a road constructed on a dead level, or with a uniform slope between points upon different levels—especially if the slope be a long one—is somewhat more fatiguing upon the draft of the horse or mule, than one with an alternation of gentle ascents and descents, of say $\frac{1}{110}$ to $\frac{1}{100}$, and that a horse can draw as heavy a load at as great a speed *up* these gradients, if they are of moderate length, as he can upon a perfect level, while in going *down* he would experience a measure of relief, hardly perceptible perhaps at the time, but which during several days of continual labor would amount to a positive benefit. This idea, although it has the appearance of great plausibility, is probably a mere popular error, unable to withstand the test of intelligent investigation. Upon a very long and steep gradient—one for instance greatly exceeding the angle of repose, and therefore inadmissible upon good roads—it would doubtless be an advantage to have short sections upon which the slope would be less than that angle, where halts could be made for rest, and the animals be entirely relieved of pressure in either direction ; but, upon a well devised road, no engineer would be justified in making special provisions for securing a succession of gentle ascents and descents, upon any considerations connected solely with the question of traction.

The proper drainage of a road requires that its side ditches should have a gentle inclination longitudinally, and, in order that the road surface may be kept free from standing water without giving it too great a rise in the middle, suitable longitudinal slopes should be given to it. For this slope English engineers generally adopt $\frac{1}{80}$, or 66 feet to the mile, and the French Corps of Ponts et Chaussees recommend $\frac{1}{125}$, or about 42 feet to the mile.

## Maximum and Minimum Grades.

As a rule therefore the gradients or longitudinal slopes of a road should be established between 1 in 30 and 1 in 125. It is generally practicable to keep within the maximum of $\frac{1}{30}$, even in locating a line upon a steep hill-side, by giving it a zigzag direction, connecting the straight portions by easy curves.

At the curves the gradients should be somewhat reduced, and the roadway made wider. The increase in width should be about one-fourth, when the angle between the straight portions is from 120° to 90°, and between one-third and one-half where the angle is from 90° to 60°. In descending a hill there is a tendency to overturn the vehicle at the curved portions, from the effects of the centrifugal force, and this danger is in proportion to the speed of the descent, and the sharpness of the turn. The radius of the curve should therefore be great, never, if practicable less than 100 feet. For the same reason, upon all sharp curves, the road surface should not be the highest in the centre, and falling in both directions so as to drain off the water, but should be the highest on the outer or convex side.

In long ascents, it is deemed advantageous to make the lowest portion comparatively steep, with a corresponding reduction in the gradient near the summit, in order that the animals may achieve, while fresh, as much of the rise as possible, while the more gentle slopes are left for the last.

## Statical Resistance on Grades.

Returning to Fig. 5 (reproduced in Fig. 6), let us suppose that the horizontal A C is of such length that the vertical rise **h**=1 foot. We then have

$$W : BC :: \mathbf{t} : 1,$$

$$\text{and } W : BC :: \mathbf{p} : AC,$$

from which we get

$$\mathbf{t} = \frac{W}{BC},$$

$$\mathbf{p} = W\frac{AC}{BC},$$

Hence, the force acting parallel to an inclined road necessary to sustain a carriage and load in a state of rest, is equal to the weight of carriage and load divided by the inclined length corresponding to a rise of one foot. And, the normal pressure of carriage and load upon an inclined road is equal to the weight of carriage and load multiplied by the quotient of the horizontal length divided by the inclined length.

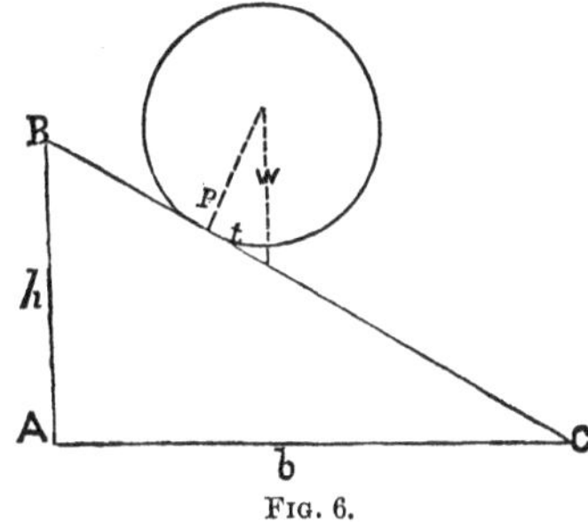

FIG. 6.

For example, the force necessary to sustain a carriage and load weighing 2500 pounds upon a road with a gradient of 1 in 20 is $\frac{2500}{\sqrt{20^2 + 1}} = 125$ pounds nearly, and the normal pressure upon the road-surface is $2500 \times \frac{20}{\sqrt{20^2 + 1}} = 2496$.

These results are theoretical. They approximate to practical correctness, as the friction on the axles is diminished, and the smoothness and hardness of the road is increased.

## Dynamical Resistances. Sir John Macniell's Formulae.

The following arbitrary formulae have been deduced by Sir John Macniell from a number of experiments upon the several descriptions of road named below, with stage-coaches and wagons moving at various velocities, and carrying various loads. R=force required to move the vehicle; W= weight of vehicle; w=weight of load, all expressed in pounds. v=velocity in feet per second. c is a constant, depending for its magnitude upon the character of the road-surface, which, for the several roads tried, is as follows:

| | | |
|---|---|---|
| On a timber surface, or on a paved road | | c = 2 |
| On a well made broken stone road | in a dry, clean state | c = 5 |
| " " " | covered with dust | c = 8 |
| " " " | wet and muddy | c = 10 |
| On a gravel or flint road, in a dry, clean state | | c = 13 |
| " " | wet and muddy | c = 32 |

For a common stage-wagon the formula is,

$$R = \frac{W+w}{93} + \frac{w}{40} + cv,$$

and for a stage-coach,

$$R = \frac{W+w}{100} + \frac{w}{40} + cv.$$

For example, the force necessary to move a stage-coach weighing 2400 pounds, loaded with 2000 pounds, at a speed of six feet per second, upon a level, dry, and clean gravel road, is $\frac{2400+2000}{100} + \frac{2000}{40} + 6 \times 13 = 172$ pounds.

## Mr. Law's Table of Dynamical Resistances.

The following table, prepared by curtailing to some ex-

| RATE OF INCLINATION. | ANGLE WITH THE HORIZON. | | | FOR A STAGE-WAGON AND LOAD OF SIX TONS, MOVING AT THREE MILES PER HOUR. | | | | FOR A STAGE-COACH AND LOAD OF THREE TONS, MOVING AT SIX MILES PER HOUR. | | | |
|---|---|---|---|---|---|---|---|---|---|---|---|
| | | | | Force required to draw the wagon UP the incline. | Force required to draw the wagon DOWN the incline. | Equivalent length of level road for an ASCENDING wagon. | Equivalent length of level road for a DESCENDING wagon. | Force required to draw the coach UP the incline. | Force required to draw the coach DOWN the incline. | Equivalent length of level road for an ASCENDING coach. | Equivalent length of level road for a DESCENDING coach. |
| | ° | ′ | ″ | lbs. | lbs. | Miles. | Miles. | lbs. | lbs. | Miles. | Miles. |
| 1 in 260 | 0 | 13 | 13 | 315 | 212 | 1.196 | .8039 | 387 | 336 | 1.071 | .9286 |
| " 250 | 0 | 13 | 45 | 317 | 210 | 1.204 | .7963 | 388 | 335 | 1.074 | .9259 |
| " 240 | 0 | 14 | 19 | 320 | 208 | 1.212 | .7876 | 390 | 334 | 1.077 | .9226 |
| " 230 | 0 | 14 | 57 | 322 | 205 | 1.222 | .7785 | 391 | 332 | 1.080 | .9192 |
| " 220 | 0 | 15 | 37 | 325 | 203 | 1.232 | .7683 | 392 | 331 | 1.084 | .9156 |
| " 200 | 0 | 17 | 11 | 331 | 197 | 1,255 | .7451 | 395 | 328 | 1.092 | .9071 |
| " 180 | 0 | 19 | 6 | 338 | 189 | 1.283 | .7171 | 399 | 324 | 1.103 | .8968 |
| " 160 | 0 | 21 | 29 | 348 | 180 | 1.319 | .6814 | 404 | 320 | 1.116 | .8839 |
| " 140 | 0 | 24 | 33 | 360 | 168 | 1.364 | .6359 | 410 | 314 | 1.132 | .8673 |
| " 130 | 0 | 26 | 27 | 367 | 160 | 1.392 | .6079 | 413 | 310 | 1.142 | .8573 |
| " 120 | 0 | 28 | 39 | 376 | 152 | 1.425 | .5752 | 418 | 306 | 1.154 | .8451 |
| " 110 | 0 | 31 | 15 | 386 | 142 | 1.451 | .5491 | 423 | 300 | 1.169 | .8308 |
| " 100 | 0 | 34 | 23 | 398 | 129 | 1.510 | .4903 | 429 | 294 | 1.185 | .8142 |
| " 95 | 0 | 36 | 11 | 405 | 122 | 1.537 | .4634 | 432 | 291 | 1.195 | .8045 |
| " 90 | 0 | 38 | 12 | 413 | 114 | 1.566 | .4338 | 436 | 287 | 1.206 | .7937 |
| " 85 | 0 | 40 | 27 | 422 | 106 | 1.600 | .4004 | 441 | 282 | 1.219 | .7801 |
| " 80 | 0 | 42 | 58 | 432 | 96 | 1.637 | .3629 | 446 | 278 | 1.232 | .7677 |
| " 75 | 0 | 45 | 51 | 443 | 85 | 1.680 | .3204 | 451 | 272 | 1.247 | .7522 |
| " 70 | 0 | 49 | 7 | 456 | 72 | 1.728 | .2719 | 457 | 266 | 1.265 | .7345 |
| " 65 | 0 | 52 | 54 | 470 | 57 | 1.784 | .2161 | 465 | 258 | 1.285 | .7143 |
| " 60 | 0 | 57 | 18 | 488 | 40 | 1.850 | .1505 | 474 | 250 | 1.309 | .6903 |
| " 55 | 1 | 2 | 30 | 508 | 19 | 1.926 | .0736 | 484 | 239 | 1.337 | .6629 |

tent a table given by Mr. Henry Law, C. E., shows with an approximation to exactness, quite sufficient to make it very valuable, the force required to draw two kinds of loaded vehicles, one weighing with its load 6 tons at a speed of 3 miles, and the other weighing with its load 3 tons at a speed of 6 miles per hour, along a Macadamized road in its usual state, with gradients varying from 1 in 7 to 1 in 600. The table also gives the length of level road equivalent to 1 mile of the inclined road, of each gradient, "that is the length which would require the same mechanical force to be expended in drawing a wagon over it, as would be necessary to draw it over a mile of the inclined road."

| Rate of Inclination. | Angle with the Horizon. | For a Stage-Wagon and load of six tons, moving at three miles per hour. | | | | For a Stage-Coach and load of three tons, moving at six miles per hour. | | | |
|---|---|---|---|---|---|---|---|---|---|
| | | Force required to draw the wagon UP the incline. | Force required to draw the wagon DOWN the incline. | Equivalent length of level road for an ASCENDING wagon. | Equivalent length of level road for a DESCENDING wagon. | Force required to draw the coach UP the incline. | Force required to draw the coach DOWN the incline. | Equivalent length of level road for an ASCENDING coach. | Equivalent length of level road for a DESCENDING coach. |
| | ° ′ ″ | lbs. | lbs. | Miles. | Miles. | lbs. | lbs. | Miles. | Miles. |
| 1 in 600 | 0 5 44 | 286 | 241 | 1.085 | .9150 | 373 | 350 | 1.030 | .9690 |
| " 550 | 0 6 15 | 288 | 239 | 1.093 | .9074 | 374 | 349 | 1.033 | .9662 |
| " 500 | 0 6 53 | 291 | 237 | 1.102 | .8979 | 375 | 348 | 1.037 | .9629 |
| " 450 | 0 7 38 | 294 | 234 | 1.113 | .8869 | 377 | 347 | 1.041 | .9588 |
| " 400 | 0 8 36 | 297 | 230 | 1.128 | .8725 | 378 | 345 | 1.046 | .9535 |
| " 350 | 0 9 49 | 302 | 225 | 1.146 | .8543 | 381 | 342 | 1.053 | .9469 |
| " 300 | 0 11 28 | 309 | 219 | 1.170 | .8301 | 384 | 339 | 1.061 | .9381 |
| " 280 | 0 12 17 | 312 | 216 | 1.182 | .8179 | 386 | 338 | 1.066 | .9336 |

| RATE OF INCLINATION. | ANGLE WITH THE HORIZON. | | | FOR A STAGE-WAGON AND LOAD OF SIX TONS, MOVING AT THREE MILES PER HOUR. | | | | FOR A STAGE-COACH AND LOAD OF THREE TONS, MOVING AT SIX MILES PER HOUR. | | | |
|---|---|---|---|---|---|---|---|---|---|---|---|
| | | | | Force required to draw the wagon UP the incline. | Force required to draw the wagon DOWN the incline. | Equivalent length of level road for an ASCENDING wagon. | Equivalent length of level road for a DESCENDING wagon. | Force required to draw the coach UP the incline. | Force required to draw the coach DOWN the incline. | Equivalent length of level road for an ASCENDING coach. | Equivalent length of level road for a DESCENDING coach. |
| | ° | ′ | ″ | lbs. | lbs. | Miles. | Miles. | lbs. | lbs. | Miles. | Miles. |
| 1 in 50 | 1 | 8 | 6 | 533 | .... | 2.019 | .... | 496 | 227 | 1.371 | .6283 |
| " 45 | 1 | 16 | 24 | 562 | .... | 2.133 | .... | 511 | 212 | 1.412 | .5871 |
| " 40 | 1 | 25 | 57 | 600 | .... | 2.274 | .... | 530 | 194 | 1.464 | .5251 |
| " 35 | 1 | 38 | 14 | 648 | .... | 2.456 | .... | 554 | 170 | 1.530 | .4690 |
| " 34 | 1 | 41 | 8 | 659 | .... | 2.499 | .... | 559 | 164 | 1.546 | .4535 |
| " 33 | 1 | 44 | 12 | 671 | .... | 2.544 | .... | 565 | 158 | 1.562 | .4370 |
| " 32 | 1 | 47 | 27 | 684 | .... | 2.593 | .... | 572 | 152 | 1.580 | .4193 |
| " 31 | 1 | 50 | 55 | 697 | .... | 2.644 | .... | 578 | 145 | 1.599 | .4007 |
| " 30 | 1 | 54 | 37 | 712 | .... | 2.699 | .... | 586 | 138 | 1.619 | .3805 |
| " 29 | 1 | 58 | 34 | 727 | .... | 2.758 | .... | 593 | 130 | 1.640 | .3592 |
| " 28 | 2 | 2 | 5 | 744 | .... | 2.820 | .... | 602 | 122 | 1.663 | .3363 |
| " 27 | 2 | 7 | 2 | 762 | .... | 2.888 | .... | 610 | 113 | 1.688 | .3119 |
| " 26 | 2 | 12 | 2 | 781 | .... | 2.960 | .... | 620 | 103 | 1.714 | .2854 |
| " 25 | 2 | 17 | 26 | 801 | .... | 3.038 | .... | 630 | 93 | 1.743 | .2566 |
| " 21 | 2 | 23 | 10 | 823 | .... | 3.120 | .... | 641 | 82 | 1.774 | .2257 |
| " 23 | 2 | 29 | 22 | 847 | .... | 3.213 | .... | 653 | 69 | 1.808 | .1919 |
| " 22 | 2 | 36 | 10 | 874 | .... | 3.313 | .... | 666 | 56 | 1.844 | .1554 |
| " 21 | 2 | 43 | 35 | 903 | .... | 3.423 | .... | 681 | 42 | 1.884 | .1150 |
| " 20 | 2 | 51 | 21 | 933 | .... | 3.538 | .... | 696 | 26 | 1.926 | .0730 |
| " 19 | 3 | 0 | 46 | 970 | .... | 3.677 | .... | 714 | 8 | 1.977 | .0221 |
| " 18 | 3 | 10 | 47 | 1009 | .... | 3.826 | .... | 734 | .... | 2.032 | .... |
| " 17 | 3 | 21 | 59 | 1053 | .... | 3.991 | .... | 756 | .... | 2.092 | .... |

| Rate of Inclination. | Angle with the Horizon. | For a Stage-Wagon and load of six tons, moving at three miles per hour. | | | | For a Stage-Coach and load of three tons, moving at six miles per hour. | | | |
|---|---|---|---|---|---|---|---|---|---|
| | | Force required to draw the wagon UP the incline. | Force required to draw the wagon DOWN the incline. | Equivalent length of level road for an ASCENDING wagon. | Equivalent length of level road for a DESCENDING wagon. | Force required to draw the coach UP the incline. | Force required to draw the coach DOWN the incline. | Equivalent length of level road for an ASCENDING coach. | Equivalent length of level road for a DESCENDING coach. |
| | ° ′ ″ | lbs. | lbs. | Miles. | Miles. | lbs. | lbs. | Miles. | Miles. |
| 1 in 16 | 3 34 35 | 1102 | .... | 4.178 | .... | 780 | .... | 2.160 | .... |
| " 15 | 3 48 51 | 1157 | .... | 4.388 | .... | 807 | .... | 2.234 | .... |
| " 14 | 4 5 14 | 1221 | .... | 4.629 | .... | 839 | .... | 2,322 | .... |
| " 13 | 4 23 56 | 1294 | .... | 4.906 | .... | 875 | .... | 2.423 | .... |
| " 12 | 4 45 49 | 1379 | .... | 5.229 | .... | 918 | .... | 2.540 | .... |
| " 11 | 5 11 40 | 1480 | .... | 5.611 | .... | 968 | .... | 2.679 | .... |
| " 10 | 5 42 58 | 1600 | .... | 6.067 | .... | 1028 | .... | 2.846 | .... |
| " 9 | 6 20 25 | 1747 | .... | 6.623 | .... | 1101 | .... | 3.048 | .... |
| " 8 | 7 7 30 | 1929 | .... | 7.315 | .... | 1192 | .... | 3.300 | .... |
| " 7 | 8 7 48 | 2162 | .... | 8.199 | .... | 1308 | .... | 3.621 | .... |

From the foregoing table we see :

1. That the force necessary to move a vehicle at a certain velocity on a level road must be decreased on a descending grade to precisely the same extent that it must be increased in ascending the same grade, in order to maintain the same velocity.

2. It must not, however, be inferred from this that the animal force expended in passing and repassing on the same road, will gain as much in descending the several grades as it will lose in ascending them. The animal force must be

adequate, either in number or in power, for achieving the steepest ascending grades on any route, and no reduction in the number of animals will be practicable in the general case, in descending that or the lower grades, or upon the level portions of the line.

# CHAPTER II.

## EARTHWORK, DRAINAGE AND TRANSVERSE FORM OF COUNTRY ROADS.

### Excavations and Embankments.

DUE regard to economy in the cost of constructing a road generally requires that its location shall be such that the cuttings shall balance the fillings, or in other words that the excavations, at points where the ground is higher than the road, shall furnish the contiguous *embankments* at points where the road is higher than the natural surface. Such, however, is not always the case, it being cheaper, under some circumstances, to deposit the excavations in *spoilbanks,* and procure the earth for embankment from *side-cuttings* near by.

The first location of the road upon the map will seldom be more than an approximation to the best line, which must finally be ascertained after successive approximations, for each of which a series of new sections must be drawn, and new calculations made. The contour lines referred to, of which an example is given in Fig. 3, or the parallel cross-sections as shown in Fig. 4, will be of great assistance in making these computations, and will materially abridge the labor of locating the line.

### The "Lead."

Prof. Mahan says, "In the calculations of the solid contents required in balancing the excavations and embankments the most accurate method consists in subdividing the different solids into others of the most simple geometrical forms as prisms, prismoids, wedges and pyramids, whose solidities are readily determined by the ordinary rules for the mensuration of solids." Other methods "consist in taking a number of equidistant profiles, and calculating the solid contents between each pair, either by multiplying the half sum of their areas by the distance between them, or by taking the profile at the middle point between each pair, and multiplying its area by the same length as before." In order to save labor and insure accuracy, tables for these calculations have been prepared and published.

Care must be taken in determining the *lead*, or the average distance to which the cuttings must be transported in making the fillings, a distance usually assumed, in each case, to be the length of the right line joining the centre of gravity of the solid of excavation and that of embankment. The least possible lead is essential from considerations of economy, and this is usually secured when such conditions are interposed that the paths over which the different portions of the solid of excavation are conveyed away to the solid of embankment, shall not cross each other either horizontally or vertically.

These conditions require that the sum of the products, obtained by multiplying all the elementary volumes by the distances over which they are respectively transported, shall be a minimum. As the computations involve the employment of the higher mathematics, they are not inserted here.

## Growth of Excavated Earth.

It must be remembered that the different kinds of earth do not fill the same volume in artificial embankments that they previously occupied in their natural bed. The *growth*, or augmentation in volume, of freshly-dug earth, varies from 15 to 25 per cent among the various kinds, but where formed and compacted into embankments, it settles or shrinks to less than its bulk in the natural bed. This shrinking for different earths, is approximately as follows: gravel or sand about 8 per cent; clay about 10 per cent; loam about 12 per cent; loose vegetable surface soil about 15 per cent; puddled clay about 25 per cent. Mr. John C. Trautwine, C. E., found that 1 cubic yard of hard rock broken into fragments made $1\frac{9}{10}$ cubic yards of loose heap; $1\frac{3}{4}$ cubic yards carelessly piled; $1\frac{6}{10}$ cubic yards carefully piled; $1\frac{1}{2}$ cubic yards of very carelessly scabbled rubble, or $1\frac{1}{4}$ cubic yards of somewhat carefully scabbled rubble.

## Moving Earth.

In excavating and removing earth, it is first loosened with picks, spades or plows, and then shoveled into the barrows or carts which convey it away. For short haulages, say of 90 to 100 feet, the ordinary road-scraper holding about $\frac{1}{10}$ of a cubic yard may be advantageously used, when the height to which the earth has to be raised does not necessitate ascents steeper than 1 in 5. For distances exceeding the sphere of scrapers, or where these cannot be advantageously employed, earth is generally conveyed in wheel-barrows. The limiting distance, when one-horse carts should replace barrows, will seldom exceed 250 feet for all the various kinds of earth. This includes loosening, loading, moving, spreading, and the wear and tear of vehicles and tools. It is stated that, upon English works, with barrows holding $\frac{1}{10}$ of a cubic yard, the limit is 300 feet.

| KIND OF VEHICLE. | LENGTH OF LEAD, OR DISTANCE MOVED IN FEET. | NUMBER OF CUBIC YARDS IN PLACE, MOVED PER DAY BY EACH VEHICLE. | PURE, STIFF CLAY, OR CEMENTED GRAVEL. | | LIGHT, SANDY SOILS. | | STRONG, HEAVY SOILS. | |
|---|---|---|---|---|---|---|---|---|
| | | | Loosened with pick. | Loosened with plow. | Loosened with pick. | Loosened with plow. | Loosened with pick. | Loosened with plow. |
| | | | cents. | cents. | cents. | cents. | cents. | cents. |
| WHEELBARROWS. | 25 | 25.7 | 14.62 | 10.12 | 8.79 | 7.52 | 11.62 | 9.12 |
| | 50 | 22.1 | 15.30 | 10.80 | 9.47 | 8.20 | 12.30 | 9.80 |
| | 75 | 19.3 | 16.02 | 11.52 | 10.19 | 8.92 | 13.02 | 10.52 |
| | 100 | 17.1 | 16.74 | 12.24 | 10.91 | 9.64 | 13.74 | 11.24 |
| | 150 | 14.0 | 18.15 | 13.65 | 12.32 | 11.05 | 15.15 | 12.65 |
| | 200 | 11.9 | 19.52 | 15.02 | 13.69 | 12.42 | 16.52 | 14.02 |
| | 250 | 10.3 | 20.95 | 16.45 | 15.12 | 13.85 | 17.95 | 15.45 |
| | 300 | 9.07 | 22.40 | 17.90 | 16.57 | 15.30 | 19.40 | 16.90 |
| | 400 | 7.36 | 25.20 | 20.70 | 19.37 | 18.10 | 22.20 | 19.70 |
| ONE-HORSE CARTS. | 300 | 28.6 | 20.98 | 15.48 | 13.50 | 12.23 | 17.98 | 15.48 |
| | 400 | 25.0 | 21.71 | 17.21 | 14.23 | 12.46 | 18.71 | 16.21 |
| | 500 | 22.2 | 22.44 | 17.94 | 14.96 | 13.69 | 19.44 | 16.94 |
| | 700 | 18.2 | 23.88 | 19.38 | 16.40 | 15.13 | 20.88 | 18.38 |
| | 1000 | 14.3 | 26.05 | 21.55 | 18.57 | 17.30 | 23.05 | 20.55 |
| | 1400 | 11.1 | 28.91 | 24.41 | 21.43 | 20.16 | 25.91 | 23.41 |
| | 2000 | 8.33 | 33.31 | 28.81 | 25.83 | 24.56 | 30.31 | 27.81 |
| | ½ mile | 6.58 | 37.95 | 33.45 | 30.47 | 29.20 | 34.95 | 32.45 |
| | 3000 | 5.88 | 40.51 | 36.01 | 33.03 | 31.76 | 37.51 | 35.01 |
| | 3500 | 5.13 | 44.11 | 39.61 | 36.63 | 35.36 | 41.11 | 36.61 |
| | 4000 | 4.54 | 47.81 | 43.31 | 40.33 | 39.06 | 44.81 | 42.31 |
| | 1 mile | 3.52 | 57.09 | 52.59 | 49.61 | 48.34 | 54.09 | 51.59 |
| | 1½ mile | 2.40 | 76.33 | 71.83 | 68.85 | 67.58 | 73.33 | 70.83 |

Beyond a certain distance two-horse wagons should take the place of carts ; and where the lead is 1¼ to 1½ miles a temporary railway and a locomotive engine and dirt-cars become desirable. Within the last few years steam excavators, capable of digging and loading 100 cubic yards per hour, at a cost of 2½ to 3 cents per yard, have been used upon some extensive works.

The cost per cubic yard of loosening, loading, transporting to different distances, and spreading soils, including repairs to vehicles and tools, is given in the foregoing table (condensed from Trautwine), based on man's labor at $1.00, horse at 75 cents, and cart at 25 cents per day ; one driver to 4 carts.

## Completed Map, and Specifications.

Having finally adjusted all questions of gradients, excavations and embankments, and re-examined the ground, the line adopted should be carefully plotted upon the map, together with longitudinal and numerous cross-sections, to show the cuttings and fillings, as well as the natural surface of the ground.

Specifications of the several kinds of work, and working drawings of bridges, culverts, etc., should also be prepared.

Upon the longitudinal section of the road, at points taken at equal intervals apart (say 50, 75 or 100 feet) the vertical distance of the natural surface above or below the road surface should be marked in feet and inches.

In order to indicate the grades, the heights of the same points above an assumed horizontal line called the *datum line*, should also be marked. These points should be numbered on the drawing.

## Fixing the Line on the Ground.

The axis of the road is located on the ground by driving stakes to correspond with the several points on the map. These stakes are lettered *cut*, or *fill*, with the number of feet added, to indicate that the natural surface at these points must be *cut down* or *filled up* the specified distance, in order to attain the position of the road surface. Stakes showing the width of the roadway, and the lateral limits of the cuttings and fillings should also be established.

*Earthwork* is a term applied to the movement and disposal of earth, whether the material handled be common earth or rock.

## Side Slopes in Cuttings.

In *excavations*, the inclination which should be given to the side slopes, and the measures, if any, which should be adopted for giving stability to their surfaces, will be governed in a great measure by the inclination and direction of the strata, the kind of soil, the degree of exposure to the action of springs, and the severity of the seasons.

On most soils, such as garden loam, and other mixtures of clay and sand, compact clay, and compact stony or gravelly soils, the slopes should be about two base to one (or at most one and a half,) perpendicular. In some cases an angle of 45°, or 1 on 1, will answer.

It is always desirable that the roadway should be exposed as fully as possible to the action of the wind and sun, in order to facilitate the evaporation of moisture from its surface. Hence, the deeper the excavation, the more gentle should be the side slopes, particularly those on the south side in high northern latitudes. When the slopes are exposed to no other causes of destruction than the action of the elements--to wind, rain, and frost—it is not essential that any special precautions should be adopted for their protection, although the expense of maintenance will be considerably lessened by sodding them, or by first covering them with three or four inches of rich soil and then seeding them down, or setting them with plants of some suitable variety of grass.

If the soil be infested with springs that might destroy the stability of the slopes, and wash them down into the roadway, they should if practicable be tapped near their source by digging into the side of the slope, and the water conveyed by blind drains or otherwise into the side drains. After the drain is constructed the earth is filled in over it, and the slope restored to the required shape. Figs. 7, 8 and 9 show cross sections of three forms of drains,

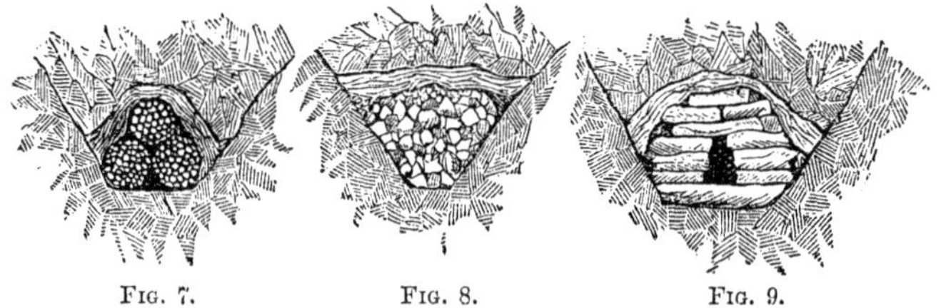

FIG. 7. FIG. 8. FIG. 9.

either of which will answer for this purpose, in ordinary cases. Fig. 7 is constructed with three fascines, laid longitudinally in the trench excavated to the source of the spring. A single fascine of equivalent area in cross-section, or loose brushwood placed compactly in the trench, may be substituted. Fig. 8 shows a blind drain of fragments of stone or loose rubble. In Fig. 9 a clear water way is left in the middle of the drain. In either case the sides and top of the drain should be covered with straw, coarse grass, small brushwood, or sods with the grass side next the drain, to prevent the latter becoming choked with earth.

In cases where the entire body of the slope is saturated with water, and the sources of the springs are unknown, or cannot be reached, good drainage may generally be secured by a series of blind drains extending a few feet into the slope at its base, or by a single drain constructed of loose stone in the form of an inclined retaining wall, parallel with the foot of the slope, as shown in Fig. 10. Similar

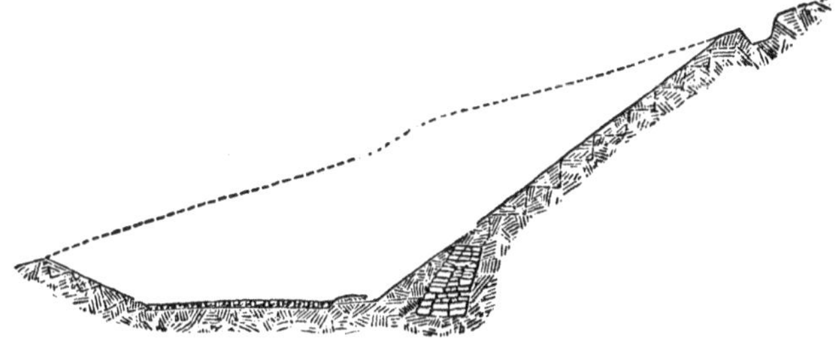

FIG. 10.

precautions to those already mentioned must be taken in this case to prevent the drain becoming choked. A series of parallel tile drains in the body of the slope, six to eight feet apart, and four to six feet below the surface, will sometimes secure good drainage in springy soils. The round tiles are the best, and the diameter of the bore need not exceed one to two inches. They are laid end to end in a narrow trench, then covered with straw, hay, or turf to prevent their choking, and the trench filled up. These drains should have an inclination of one in eighty to one in one hundred, and may all empty into a masonry conduit or earthen pipe leading down the slope into the side drains of the roadway. Instances may occur where a deep narrow trench dug just beyond the crest of the side slope, and filled up with broken stone or pebbles, will cut off all the springs. As such an arrangement will also prevent the surface water from higher levels from running over the slope, it may be the least costly method of effecting good drainage. Deep shafts for collecting the water are sometimes sunk from the natural surface above the crest of the slope, and the water conveyed from them into the side drains of the roadway, through pipes or drains laid by tunneling. In deep cuttings in clayey soils, of such character that water will render them sufficiently plastic to slide down the slope, or soils easily cut into gullies by running water, it is well to form the slope in benches or *berms*, shaped into shallow drains called catchwater drains, to receive the water and earth from the higher levels. They should have a slight inclination longitudinally, so as to discharge their water at the foot of the slope, or, at suitable intervals, into paved open drains running directly down the face of the slope.

In soils formed in great part of unctuous clay, or alternate layers of clay and sand, there is always a tendency to *slides* during the wet season, or during the spring in high latitudes, when the frost is leaving the ground. Under these circumstances special precautions should be taken for the protection of the side slopes, such as (1) cutting off all springs if any exist, (2) turning off the water from the higher levels by a drain above the crest of the slope, and (3) arrangements for conveying off the water which falls upon the slopes, either by carefully constructed catchwater drains, by sodding the slope, or by a combination of the two. Before top-dressing the slope with rich soil, preparatory to sodding or seeding down, it should be cut into horizontal benches or steps to guard against slides, as shown on the right in Fig. 10. This method may be advantageously applied to those kinds of slaty rocks which disintegrate rapidly by alternate freezing and thawing. (The benches are incorrectly represented on the left hand slope in the figure.)

Although the sides of a cutting through compact rock would stand firmly, if left in a vertical position, it is desirable that they should be sloped to some extent so as to expose the road-surface to the drying influences of the wind and sun. Unless prevented by considerations of economy, the slope on the side next the equator in high latitudes, should be as great as one base to one perpendicular, while one base to two perpendiculars will ordinarily answer on the opposite side.

When the depth of the excavation exceeds a certain limit, generally assumed to be about 60 feet, it is in most cases cheaper to tunnel than to make a cut. It is however, seldom necessary to resort to this expedient, in constructing common roads, for which indeed it is peculiarly inappropriate, on account of excluding the wind and sun.

## Embankments.

Embankments should be made sufficiently firm and compact to resist all tendency to unequal settlement and slides. They should therefore possess not only great but *uniform* solidity, especially if they are high, conditions which are best secured by forming them in successive horizontal layers well compacted by ramming. As this method is expensive, it is usual, from considerations of economy, to carry out an embankment to its full height from the beginning, at the same time making the cut, which supplies the earth, to its full depth. The earth, on being tipped at the end of the bank undergoing formation, slides down the slope and finally comes to temporary rest, approximately at the angle of repose. (See Fig. 11).

As the rapidity with which this kind of work can be executed depends upon the number of "tipping places"

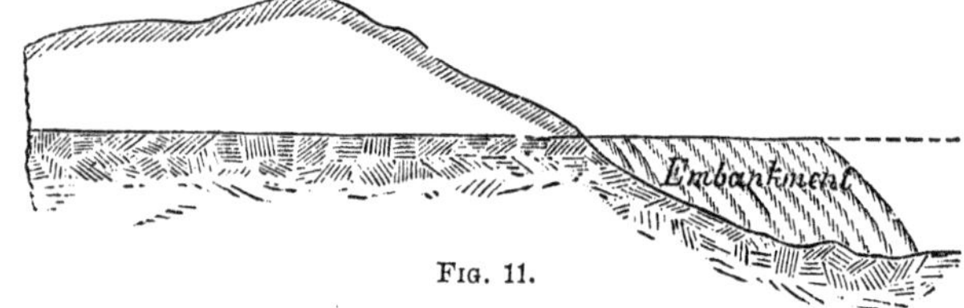


FIG. 11.

afforded by the width of the embankment, it is usual to form the latter at first broader at the top, and correspondingly narrower at the bottom than the required dimensions, maintaining of course the requisite area of cross-section. The excess at the top (the angles at A and C, Fig. 12) is after-

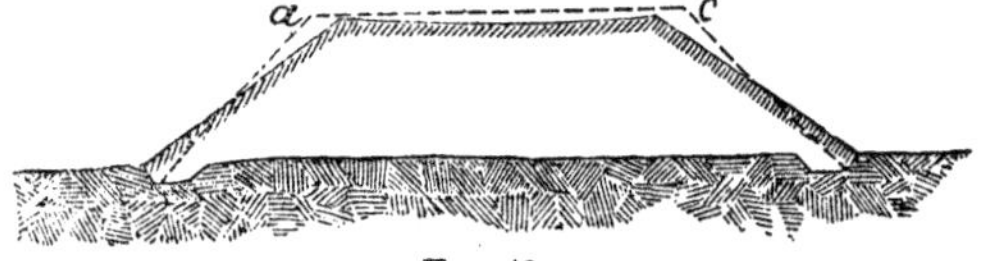


FIG. 12.

wards moved down to the bottom, thus securing the required width of base, and inclination of side slopes.

The sides of the embankment should always be kept somewhat higher than the centre line, in order to retain the rain fall, and consequently hasten the consolidation of the mass. For the same reason when the case will warrant the expense of constructing in successive horizontal layers they

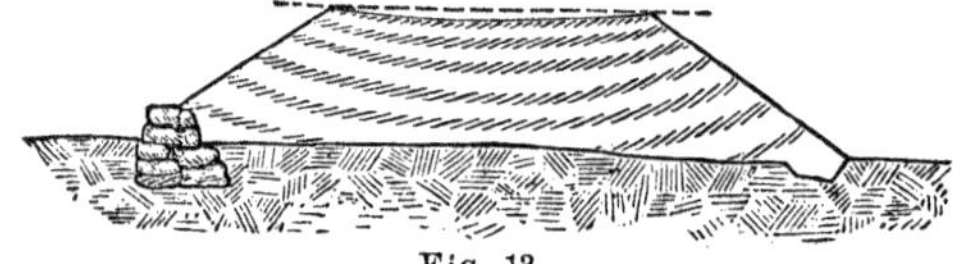

FIG. 13.

should be made concave on the top, as shown in Fig. 13. This will also lessen the danger of slides.

## Embankment Slopes.

The foot of the slope may be secured by resting it in a shallow trench, or by abutting it against a low dry wall of stone (Fig. 13), and, in localities where there is an abundance of stone to be obtained in the proper shape at moderate cost, the entire slope may sometimes be advantageously replaced by a sustaining wall laid up without mortar. In cases where the embankments require more earth than the necessary excavations afford, a sustaining wall may be cheaper even in its first cost, than a slope of earth, while the current cost of its maintenance will be small in comparison with that of earthwork.

The inclination of the slopes should be less than the earth will naturally assume, in order to give them greater stability, and they should be protected by sodding or seeding down.

The surface water of the top may be collected by side-drains and carried down the slopes at intervals either in paved gutters or blind drains. If allowed to shed itself over the slope, gullies would be formed, and the embankment eventually destroyed.

When extensive excavations have to be made through rock, and stone for embankment purposes is therefore plenty and cheap, the entire face of the slope may be roughly paved or covered with stone, at moderate cost, rendering all other precautions for its preservation unnecessary. Fragments of various sizes and shapes, from one to two feet in length, may be used for this purpose. They should be placed generally with the longest edges down, their faces at right angles to the slope and parallel with the axis of the road. It is not usual to protect the slopes of embankments in this careful manner, but they should never be required to carry off the surface water of the roadway.

### Hill-Side Roads.

A roadway located upon a hill-side is usually formed half in excavation, and half in embankment, allowance being made for the shrinkage of the latter. To guard against slides, the natural surface should be prepared to receive the embankment, by cutting it into benches as in Fig. 14. The

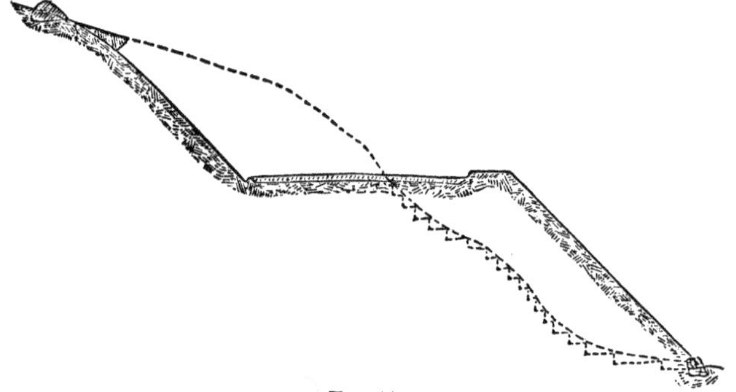

FIG. 14.

foot of the slope should abut against a low dry stone wall below the reach of frost. At the crest of the slope in excavation, an open trench should be formed to intercept and convey away the surface water from the higher ground.

Upon steep hill-sides, the side slopes of excavation and of embankment must both be replaced by sustaining walls. Dry walls will usually answer for these purposes, if the stone can be procured of suitable sizes. (Fig. 15.)

If the hill-side upon which the road is located be a rocky ledge of less inclination than one perpendicular to one base, the same method of construction by making the excavations supply the embankments may be followed. The enrockment *filling* in this case will occupy a greater volume than the cutting from which it is taken.

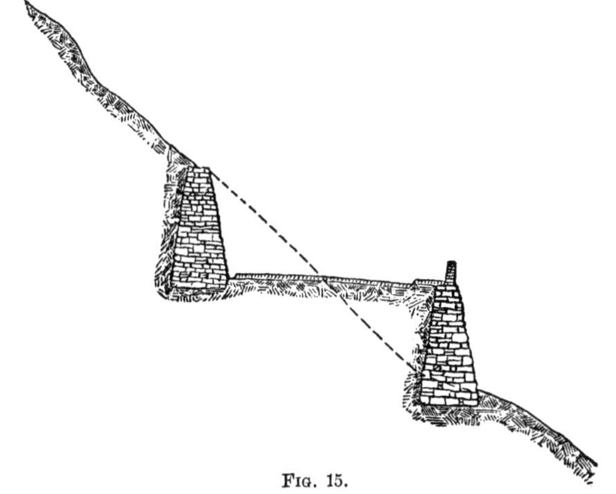

Fig. 15.

When the natural slope of the ledge is very steep, as for example when it is steeper than one base to one and a half perpendicular, the whole roadway may be formed in excavation, or, as shown in Fig. 16, by cutting the face of the ledge into two or more horizontal steps with vertical faces, and building up the embankment in the form of a solid stone wall, in horizontal courses, either with or without mortar. In the figure the lower step, on which the wall rests, may sometimes be advantageously replaced by two smaller steps, *a, c,* and *b, b.* On account of the great comparative cost of excavations in rock, estimates for work of this character should always be based upon numerous and careful sections. All attempts to lessen the quantity of excavation by increasing the number and diminishing the width of the steps, require additional precautions against settlement in the built-up portion of the roadway.

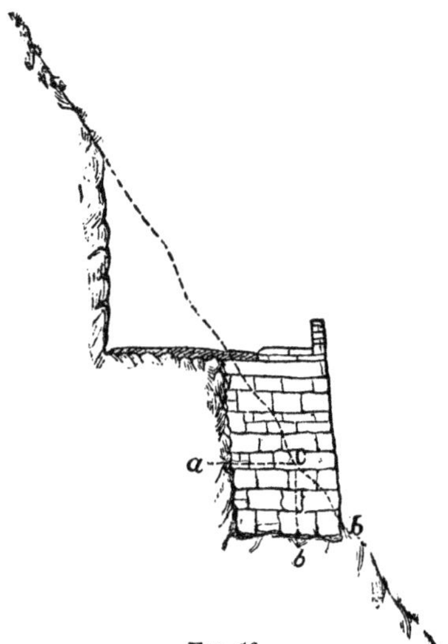


FIG. 16.

## Roads over Marshes and Swamps.

If the road runs through a swamp or marsh resting upon a firm substratum, the soft material should be removed when its depth does not exceed two or three feet, and the road embankment formed directly upon the hard soil. A wide and deep open ditch should be cut in the marsh on each side of the road, to receive the surface drainage and cut off the water from the adjacent marsh.

Roads over deep marshes must be constructed upon a different plan. A system of thorough deep drainage being essential, the method usually followed is to cut a deep and wide ditch on each side, leaving between each ditch and the road covering an unoccupied strip or berm several feet in width. Cross drains should connect the two main drains at frequent intervals, to drain the soil under the roadway. These cross drains are formed by digging trenches a little deeper than the lowest water level in the side ditches, and filling them with fragments of stone. If water permanently stands in the main drains, the cross drains may be filled below the water level with brushwood or fascines. On the foundation thus formed the roadway may be constructed.

In deep marshy soils having a spongy subsoil, it is sometimes necessary to form an artificial bed for the road covering. In some cases of rather firm marsh, it is sufficient to remove the soft top-soil to a depth of three or four feet, and substitute for it sand, gravel, or other compact material that will not retain water. Upon this bed the road covering is placed. In others a bed formed of fascines has been used with success. Fascines are made by binding together, by wires or withes, in cylindrical bundles nine to ten inches in diameter and ten to twenty feet long, slender branches of underwood. A layer of fascines, placed across the road, side by side, is first laid down. A second layer at right angles to these follows, and if necessary a third transverse layer, and so on until the required height of road bed is attained. Stout stakes or pickets are driven through the entire thickness of fascines to keep them firmly in place, or, the withes may be cut after each layer is in position, so as to allow the brushwood to assume the form of a layer of uniform compactness. The top layer should be placed transversely to the line of the road. Having prepared the foundation in this manner the road covering is placed upon it in the usual way. If deemed necessary, cross blind-drains leading to the side ditches, may be introduced under the bed of fascines in order to secure deeper sub-drainage.

## Roads over Tidal Marshes.

Roads constructed over marshes which are subject to

daily overflow from the tides, may be effectively drained, down to the level of ordinary low water, by a very simple system of sluice-gates. The ordinary case is where the road crosses a tidal stream and then traverses the marsh bordering thereon. Under such circumstances, the earth excavated in cutting the two parallel side ditches, should be formed into a dike or levee on the outer side of each ditch and high enough to exclude high water. The ends of these dikes should be connected, along the margin of the stream where the road joins the bridge, by a dike of the same height, thus surrounding the roadway and side ditches with a continuous impervious bank of earth, rising above high water level.

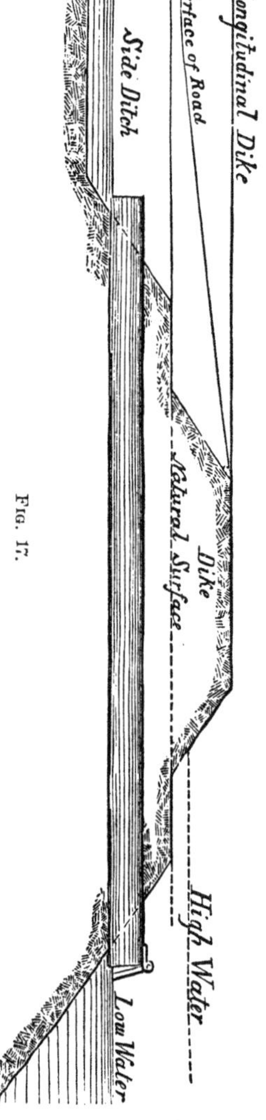

Fig. 17.

Pipes of wood or iron, provided with valves on the outer end opening outward, are then inserted, at the level of low tide, through the dike which separates the ends of the side ditches from the tidal stream. The water flows out through these pipes whenever the tide level outside is below the water level in the ditches. When the reverse ensues the gates are closed by the external pressure of the water, and all inward flow prevented. Hence the water in the ditches remains at the level of low tide, and the surface of the road should be established at such height above that level that it will always be firm and solid. Whatever level may be adopted for the roadway across the marsh, it must on approaching the stream rise to the top of the dike to prevent overflow from the stream at that point. This point is illustrated in Fig. 17, which is a longitudinal section through the sluice-pipe. The road is built a little higher than the natural surface, and ascends to the crest of the dike, near the stream. The bridge across the stream in continuation of the road is not shown.

This method of drainage has been very successfully applied in reclaiming the "Jersey Flats" between the cities of Newark and Jersey City, New Jersey, a large tract of marsh land with its natural surface not higher than the level of ordinary high tide, and in large part some inches below that level. These Flats were formerly subject to twice daily overflow, from the Passaic and Hackensack rivers, in which the mean rise and fall of the tide is about 4 feet. The tract was surrounded by a dike or levee formed along the margin of the stream, of the material excavated from a wide open ditch parallel to it on the inside. Numerous broad cross ditches, dividing the marsh into parallelograms, lead into this main ditch. The sluices are located at suitable intervals so as to drain from the main ditch into both the Hackensack and the Passaic rivers. The pipes are usually placed in pairs, and are of various sizes, in only a few instances exceeding two feet in interior diameter. It has been found that a difference of one inch in the inside and outside water level, will open or close the sluice-gates.

By this means a noxious swamp has been converted into arable and thrifty land. It is sufficiently firm to support a roadway upon its surface, without any special precautions in preparing the road bed.

When the embankment requires more earth than the excavation can supply, the deficiency is made up from *side-cuttings*, made in some convenient locality near by. On the other hand when the embankments do not consume all the earth furnished by the excavations, the excess is deposited in *spoil-banks*, usually located on a somewhat lower level than the road surface, care being taken to provide for the drainage in such a manner as to prevent the formation of pools of water that might affect the stability of the side slopes of the excavations.

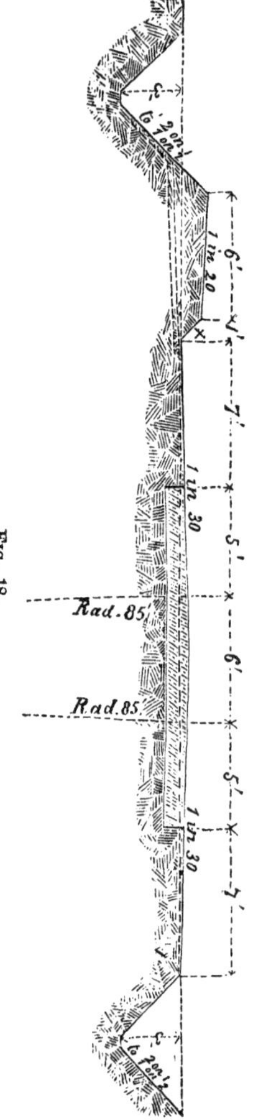


Fig. 18.

## Side Drains.

It is essential to the proper condition of a road, that the surface water of the soil adjacent thereto shall be cut off by suitable open side drains, so that it can not filter under the roadway, and render the subsoil soft, spongy, and incapable of sustaining the weight upon it. Otherwise the road covering would sink into the soft earth beneath, and require to be frequently renewed. In flat and level countries, the side drains (Fig. 18), should be at least two and a half to three feet lower than the bottom of the road covering, and, to prevent accidents to vehicles, they should be placed on the field side of the fences or hedges, when passing through sections of country where such barriers are necessary, as they usually are in agricultural districts. It would be better for the roadway, to place the side ditches between the road and the fences, thus widening the space between the latter, and in-

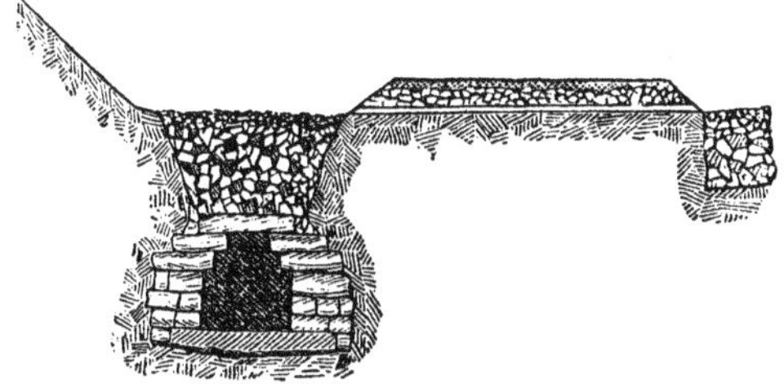

Fig. 19.

creasing the degree of exposure of the road surface to the action of the wind and sun ; and this method should be followed whenever admissible. A line of hedge or shade trees on each side of the road, exerts a very damaging effect upon its condition, and adds greatly to the cost of its proper maintenance. High walls and hedges are more objectionable than open post-and-rail or rail fences.

Upon those portions of the road in excavations, unless it be through rock, open side ditches are inadmissible, as they would soon be filled up by the earth washed down from the side slopes. Covered side drains are necessary. They may be constructed as shown in Fig. 19, with a flooring of concrete, flagging stones, or brick, with side walls of the same material, and covered with flagging stones or with bricks, or stones corbeled out to meet above the centre of the drain. The roof should be laid with open joints, and then covered with a layer of straw, hay, or fine brushwood, upon which a filling of fragments of stone, bricks, or coarse gravel and pebbles is laid, so as to allow the water to filter freely through, without carrying sediment with it.

## Cross Drains.

Besides the covered side drains in cuttings, cross drains are usually deemed necessary to keep the road bed dry. Their depth should be 20 to 24 inches below the road covering. They should have an inclination on the bottom, from the axis of the road to the side drains, of not less than 1 in 100 nor greater than 1 in 30. When the road is level they may run straight across. When otherwise, their plan assumes the form of a broad letter V, with the point in the centre of the road directed toward the ascent. From this, their usual form, they are termed *cross-mitre* drains. Their distances apart will depend upon the nature of the soil, and the kind of road covering used. In some cases, it should not exceed 18 to 20 feet, in others it may be much greater. They are constructed by digging trenches across the road bed, after the surface has been prepared for the reception of the road covering, and then filling the trench with broken stone or pebbles, leaving a small open water-way at the bottom, so constructed that it will not become choked with earth. Bricks may be used for this purpose in a variety of ways as shown in Figs. 20, 21, 22 and 23. Flat stones may be used as shown in Fig. 9, page 41.

The area of the water-way of Fig. 20, is unnecessarily large, for most cases that will occur in practice, being 16 square inches. The bottom of the road material rests on the line AB. Small drains placed close together will drain the

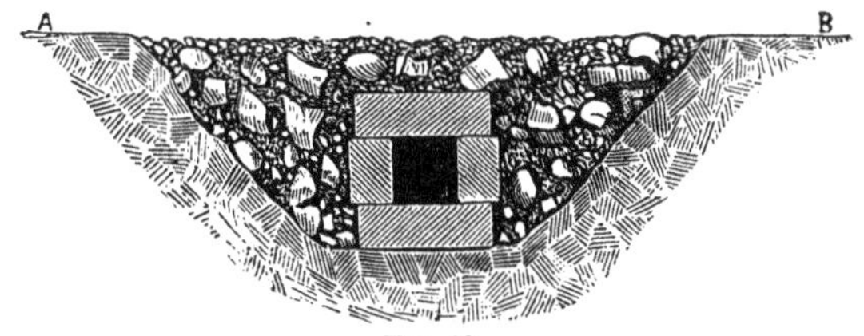

Fig. 20.

road bed much more efficiently than larger ones of the same aggregate water-way placed farther apart. A water-way of 2.5 to 4 inches sectional area, such as may be obtained with

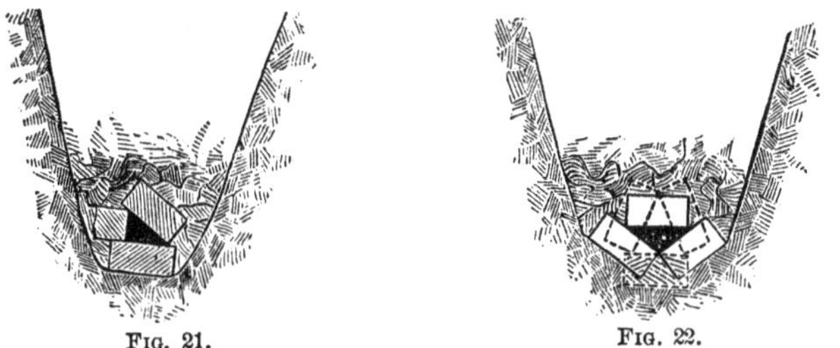
Fig. 21. Fig. 22.

three lines of whole bricks placed end to end as in Fig 22, will generally be ample. The full lines show one method of

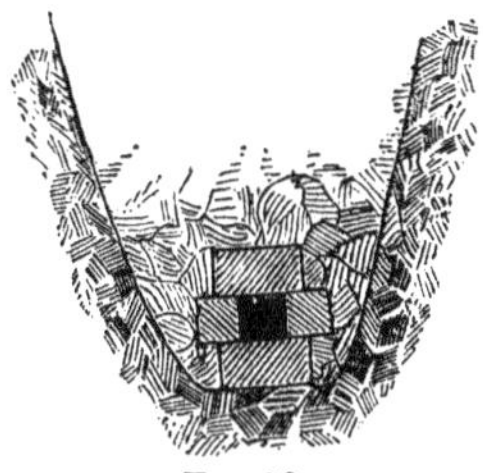
Fig. 23.

arranging the brick, and the dotted lines another. By splitting one-third of the bricks into halves longitudinally, they may be placed as in Fig. 23, or more economically still as in Fig. 21, by splitting one-fifth of them.

The water-way may be formed of drain tiles of 1½ to 2 inches interior diameter. Indeed, the ordinary method of tile-drainage for agricultural purposes will answer excellently for the sub-drainage of roads, and it will seldom be necessary to use tiles of larger bore than 1½ to 2 inches. With a fall of 1 foot in 100, a 1½ inch tile will discharge nearly 12,000 gallons of water, and a 2 inch tile nearly 22,890 gallons in 24 hours. The tiles are placed in contact, end to end, in a trench cut very narrow at the bottom, care being taken to give each piece a firm bed, and to arrange the axes in a continuous line, so as not to diminish the water-way by jogs at the joints. The bricks, stones or tiles used to form the water-way, should be covered over with a layer of hay, straw, tan-bark, turf, or other suitable material, to prevent earth from entering the drain ; the trench is then filled up with the earth excavated from it.

### Cost of Stone, Brick and Tile Drains.

Tiles are generally a little more than one foot in length, so that making a fair allowance for breakage and imperfect pieces 1000 tiles may be relied upon to lay 1000 feet of drain. Two-inch tiles can be manufactured at profit in ordinary times for $11.00 to $11.50 per thousand, and bricks for $6.50 to $7.00 per thousand. We will estimate the tiles at $14.00 and the bricks at $8.00 delivered and distributed along the road. A cross drain under a road 30 feet wide, with one foot-path, page 60, will be about 45 feet long. The tiles will cost $0.63 for each cross drain, the brick $1.35 for the small triangular drain, Fig. 21 ; $1.62 for the triangular drain Fig. 22, or the small square drain, Fig. 23, and $3.24 for the large square drain, Fig. 20. The labor will be nearly proportional to the amount of excavation, and will therefore be the least for the tile drain, so that it is within limits to estimate the cost of the latter in labor and material, at considerably less than one-half that of the former. If the cross drains have their discharge three inches above the bottom of the side ditches, with a fall from the centre of the roadway each way of three inches—equal to 1 in 90—their average depth below the road surface will be about three feet, and the average depth of excavation below the subgrade prepared for the reception of the road materials will be a little over two feet, inclusive of the greater depth under the foot-path. The following estimates of the cost per rod of these several kinds of drains, is believed to be fair, with labor at $1.75 per day, in stiff clay soils :

| | Tile drain. 2-inch pipe tiles. | Brick triangular drain. Fig. 21. | Brick drain. 3 courses end to end. Figs. 22, 23. | Brick drain. 4-in. × 4-in. Fig. 20. |
|---|---|---|---|---|
| Cutting and filling | $ cts. | $ cts. | $ cts. | $ cts. |
| per rod.......... | .25 | .30 | . 30 | . 35 |
| Cost of tiles, or bricks | .23 | .55 | .594 | 1.188 |
| Total cost per rod... | .48 | .85 | .894 | 1.538 |

The trenches for stone drains must be excavated to a width of at least 21 inches on the bottom. They may be cut with vertical sides. At least one-half of the earth must be hauled away to make room for the stone, so that there will be more than one cubic yard of earth to be carried away and a like amount of stone collected and brought to the road for every rod of drain. If stones of the right kind are near at hand it would not cost more than 25 cents per cubic yard to collect them, while twice that amount would not be excessive if they have to be brought from a distance or dug from a bank or pit. Assuming that the stones can be collected within a quarter of a mile of the road, the following estimate is submitted as a near approximation:

## Cost of Stone Drains per lineal rod.

| | |
|---|---|
| Cutting trench and hauling away the surplus earth......... | 50 cents. |
| Collecting and hauling 1 cubic yard of stone............... | 25 " |
| Laying the stone and filling in........................ | 25 " |
| Total per lineal rod..... | $1.00 |

A standard work on Farm Drainage (Henry F. French) estimates that with tiles at $10.00 per M,—1 cent per foot—and labor at $1.00 per day, the cost of a tile drain 4 feet deep will be for cutting and filling 33⅓ cents, and for the tiles 16⅔ cents, or a total of 50 cents per rod. At the same price for labor, the cost of a stone drain is set down at $1.25 per rod, viz., for cutting and filling trench 21 inches wide 50 cents, for hauling stone 50 cents, and for laying the same 25 cents. And the conclusion is that "drainage with tiles will generally cost less than one-half the expense of drainage with stones, and be incomparably more satisfactory in the end."

As tile drains are more liable to injury from frost than those of either bricks or stones, their ends at the side ditches should not, in very cold climates, be exposed directly to the weather, but may terminate in blind drains, reaching under the road a distance of about 3 to 4 feet from the inner slope of the ditch.

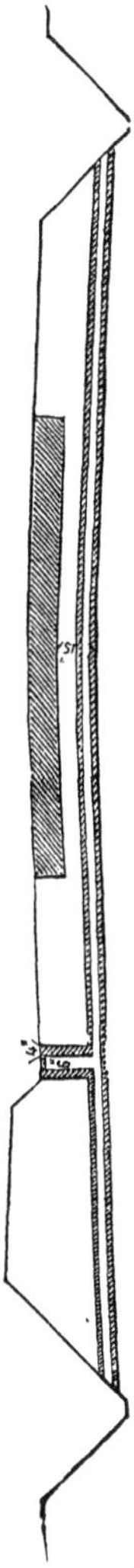

Another method of draining the road-bed, offering security from frost, is by one or more longitudinal drains, discharging into cross-drains placed from 250 to 300 feet apart, more or less, depending on the contour of the ground. With a roadway and foot path 45 feet wide at the level of the bottom of the side ditches, two such drains of 1½ to 2 inch tiles will be required in most kinds of clayey soils. They should be placed at equal distances from the side ditches and from each other, and will therefore be 15 feet apart. Their depth below the surface should be not less than 3 to 3½ feet, and the cross-drains into which they discharge should be of ample dimensions.

FIG. 24.

All roads upon clayey soils in flat level countries should be amply provided with drains under the road covering. For deep, marshy, soils, as already mentioned, they are indispensable.

### Surface Drainage.

The drainage of the road-surface is provided for by making it a few inches higher in the centre than at the sides, sloping it gently in both directions to the side gutter, and is also greatly facilitated by giving the surface an inclination longitudinally not greater than 1 in 34 nor less than 1 in 125. A series of gentle undulations may be established, even in a perfectly level country, without adding materially to the cost of construction.

If there are no sidewalks the surface drainage is discharged directly into the side ditches. When there are sidewalks the drainage is into the side gutters, from which it must be carried into the side-ditches by small covered drains of stone, brick, or tile, see Fig. 25, and dotted lines, Fig. 18, or conducted by vertical shafts into the cross-drains, if that method of draining the road-bed has been

FIG. 25.

adopted. These vertical drains (Fig. 24) are covered at top by a grating to arrest coarse earth, leaves, etc., and may have beneath them a small vault lined with masonry, termed a silt basin, to collect the fine sediment which flows in from the roadway, which should be removed from time to time and restored to the road surface.

The covered side-drains (Fig 19) may if desirable, be constructed with a roof of concrete or of brick or stone laid in mortar, in which case the filling of broken stone, gravel, etc., above, for filtering purposes, would be omitted, and arrangements made for conducting the water into the drain through vertical shafts placed at suitable intervals. These vertical openings should be placed over *silt-basins,* and be large enough to admit a boy, so that the sediment can be removed as often as necessary. (Fig. 25.) This method of construction is much used in cities and towns, but not on country roads.

## Culverts.

Besides the *side gutters, side drains, cross drains,* and *small drains* already mentioned, which are intended more especially to carry off the water falling upon the roadway, and upon the side slopes in cuttings, and to keep the road bed and subsoil dry in flat countries where the soil is clayey or marshy, other drains called *culverts* are required for carrying under the roadway the stream intersected by it, and generally for conveying away into the natural water-courses the water collected by the side gutters and ditches on the upper side of the road.

The dimensions of the water-way of culverts should be proportioned to the greatest volume of water which they may ever be required to carry off, and should in all cases be large enough to allow a boy to enter for the purpose of cleaning them out. Eighteen inches square, or if circular, twenty inches in diameter, will suffice for this purpose. They should have an inclination on the bottom of not less than one in one hundred and twenty, nor greater than one in thirty. Small culverts may be constructed of the same cross-section, and in substantially the same manner as the covered side drains, Fig. 19, with the exception that no arrangement need be made for receiving the water from above through a filter of stone fragments or gravel. The sides and roof may therefore be laid in mortar, and the floor had better be in the form of an inverted arch.

In localities where stone and brick are expensive, small culverts may be constructed of four slabs or planks (Fig. 26) forming a long box open at both ends. To prevent the side pieces from being forced together by the pressure of the surrounding earth, they should rest against small blocks of wood nailed at intervals into triangular notches cut on the inner faces of the top and bottom pieces, or the edges of the side pieces may be inserted into longitudinal grooves cut in the top and bottom pieces.

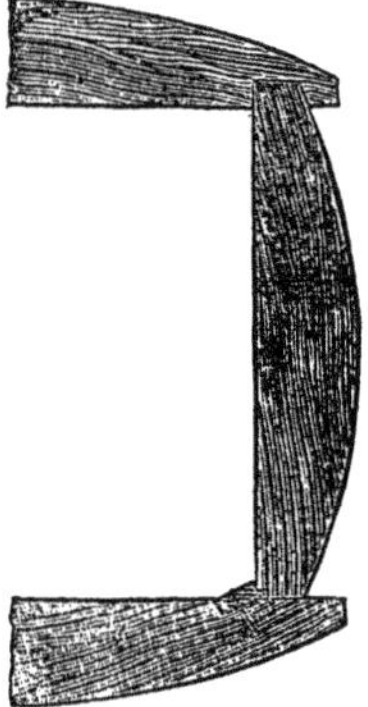
FIG. 26.

It will be found advantageous to use hydraulic concrete for culverts, especially for those of large dimensions, unless bricks can be procured at a low cost, or stone in suitable form is plenty in the neighborhood. Most localities will furnish sand, coarse gravel and pebbles. With these and a liberal addition of common lime to the hydraulic cement, a concrete suitable for work of this description can be prepared at moderate cost. The following proportions will answer:

1 measure of Rosendale, (or any equivalent) cement.

1 measure of slaked lime in powder.

4 to 4½ measures of clean sharp sand.

9 to 10 measures of pebbles, small fragments of stone or brick, oyster shells, or a mixture of them all.

When Portland cement of standard quality is used, the proportions may be:

1 barrel Portland cement, as packed for market.

1 barrel common lime, producing $2\frac{1}{4}$ to $2\frac{5}{8}$ barrels slaked lime powder.

9 to 10 barrels sand.

15 to 17 barrels of the coarse materials.

It must be borne in mind that the addition of common lime weakens, and also retards the induration of the concrete. The quantity added must, therefore, be governed by the importance of the work, and the length of time it will have to harden before being subjected to heavy weight or pressure.

A form of cross-section for culverts, which admits a min-

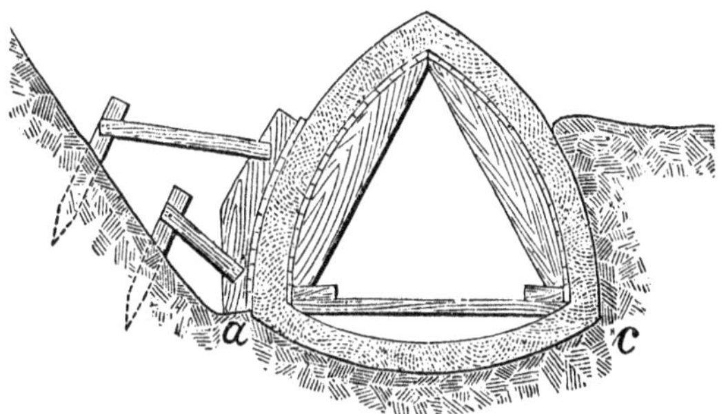


FIG. 27.

imum thickness in the concrete floor and sides is shown in Fig. 27. In constructing a culvert of this form, a trench, concave at bottom, is first excavated to the width of the base *a c*, and the concrete floor, four to five inches in thickness, is formed thereon in one layer. Upon the floor cross pieces are placed at intervals of a few feet, to support the two detached segments of centering for the sides, which should rest upon wedges, to facilitate their removal for further use. The centering may be made in lengths of from twelve to twenty feet, and of such weight that three or four men can easily handle them. Convexity is given to the exterior surfaces by a movable form, made by nailing narrow strips of boards to cross pieces cut to the required concavity. It is kept in place by wooden braces nailed to stakes driven into the side slopes of the trench. For a culvert of twenty-four inches interior width at the base, arched as in Fig. 27, with circular segments whose centres are in the vertices of the opposite angles, the sides need not exceed five to six inches in thickness, in case of ordinary depth.

For culverts with a gentle inclination on the bottom, the concrete floor may generally be safely replaced by a layer three to four inches in thickness of broken stone or pebbles of half an inch to two inches in diameter, grouted with hydraulic cement to keep them in place, and protect the foundation from undermining. For the same reason, the foundations should be started lower than when the bottom of the water way is well paved, or when formed of a monolith of concrete in the manner last described. For the discharge of large volumes of water, the masonry must be proportionally massive, but may still be made of concrete, which may be considerably cheapened by embedding in it, as the work progresses, fragments of rock of various shapes and sizes. A section of such a work is shown at Fig. 28.

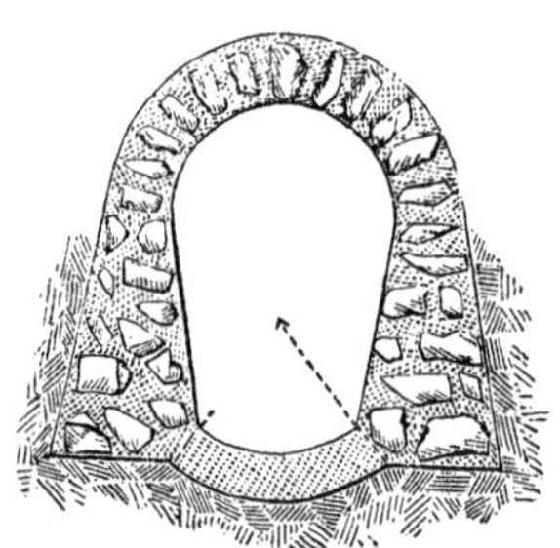
Fig. 28.

The ends of culverts passing through embankments should be protected against the undermining action of the water. This may be done by a sheet piling of flagging-stones or stout planks, sunk well into the soil, or, by an apron of rip-rap stone, or a good pavement, so as to prevent all percolation of water under or at the sides of the culvert. The length of the covered portion of a culvert is equal to the distance through the embankment, on a line with the crown of the arch or roof, and it should be extended out and finished at each end by two wing walls spread out in fan shape, and finished on top in steps, by courses or in a surface parallel with the side slope, but rising a few inches above it, to prevent the earth washing over.

In order to give greater stability to the wing walls, and increase the power to resist the pressure of the earth behind them, their plan may be that of an arc, with either the convexity or concavity to the front, the object being simply to increase the moments of inertia without materially increasing the amount of masonry in the wall. One end of the culvert given in transverse section in Fig. 28, is shown with wing walls in longitudinal vertical section and elevation in Fig. 29, and in horizontal section and plan in Fig. 30.

## Width and Transverse Form of Roads.

The determination of the width and transverse form of a country road presents questions of great importance. Some engineers recommend narrow roads, on the erroneous presumption that the cost of maintenance, like that of construction, varies directly or nearly so with the area of the road surface, while in point of fact, unless in special and extreme cases, it varies with the amount of traffic upon it, increasing, however, more rapidly than the traffic. It may be assumed that the quantity of material required to repair a road is about the same, for the same amount of traffic, whether the road be twenty-five feet or thirty feet in width, although there is a small saving in the labor of spreading it, in favor of the narrow road. A narrow road is less exposed

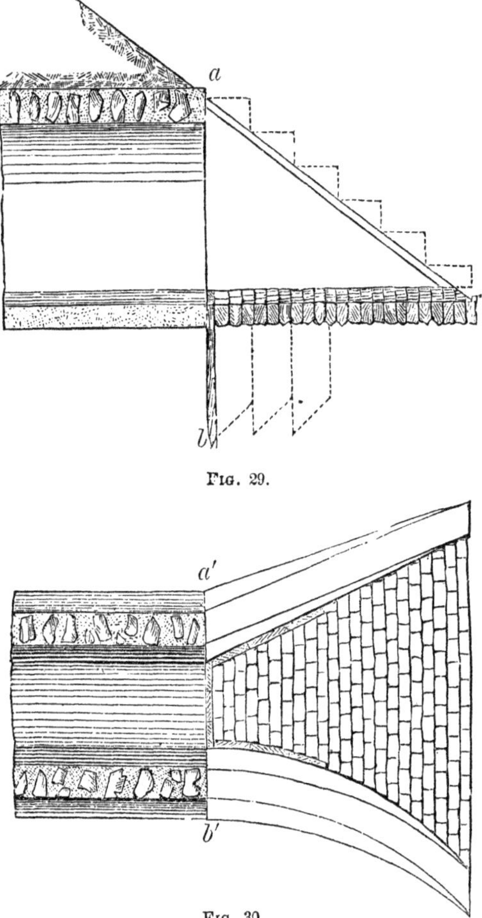


Fig. 29.

Fig. 30.

to the drying action of the wind and sun than a wide one, and also requires more constant supervision and more frequent repair, in consequence of the traffic being more closely restricted to one track.

A width of 27 to 30 feet, prepared for vehicles, will be ample for the principal thoroughfare between cities and large towns, which should be increased, within or near the cities, to 40, 50, or even to 60 feet, where the amount of traffic is large, and there is a great deal of light travel and pleasure driving.

For cross, branch, and ordinary town and country roads, the width of the portion bedded with stone may usually be reduced to from 16½ to 17 feet, which will amply suffice for two carriages of the widest usual size to pass each other upon the road-covering without danger of collision. In special cases of private roads where the bulk of heavy traffic is all in one direction, the width of covering may be restricted to 8 or 9 feet, or what is sufficient for one carriage, the loaded vehicles having the right of way, and requiring those travelling light to turn out upon the sides.

*Side-walks* for foot passengers are usually omitted in new or thinly settled countries, although always desirable, if for no other reason than the protection they afford against upsetting into the side ditches during the night, in localities where there are no fences, or where the ditches are between the fences and roadway.

At least one paved or otherwise properly covered side-walk is necessary near towns and villages. When the natural soil is composed principally of sand and gravel it may form the surface of the side-walk, which should be established at about the height of the centre of the roadway. In heavy clayey or loamy soils an excavation to the depth of five or six inches should be made to the proper width of the foot-path, and filled in with coarse sand or gravel, or with a layer of well-compacted broken stone topped off with one to two inches of gravel. The inner edge of the side-walk should be protected against the wash of the side-gutters by a facing of sods or dry stone.

In cities the side-walks are paved, and the surface slopes toward the street at the rate of not less than one inch in ten feet, in order to secure the prompt discharge of the surface water into the side-gutters, and the edge next the gutters is faced with slabs of stone called *curb-stones* set on edge, with their top edges flush with the side-walk pavement, and their lower edge six to eight inches below the bottom of the gutter.

(The construction of side-walks will be described hereafter).

In France, four classes of roads are prescribed as follows : *First*, 66 feet wide of which 22 feet in the middle are paved or stoned. *Second*, 52 feet wide of which 20 feet are stoned. *Third*, 33 wide of which 16 feet in the middle are stoned ; and *fourth*, a width of 26 feet of which 16 feet in the middle are stoned.

Telford's Hollyhead road, which runs through a hilly country, is 32 feet wide between the fences on flat ground, 28 feet on side cuttings not exceeding three feet deep, and 22 feet along steep and precipitous ground.

The Cumberland or National road in the United States has a prescribed width of 80 feet, but the prepared roadway is only 30 feet wide.

The Roman Military roads were narrow, being only 12 feet wide on the straight portions, and 16 feet upon curves.

Wide roads are sometimes finished with a road-covering in the middle only, of just sufficient width for the vehi cles to pass each other upon it, while the sides are maintained as *dirt roads,* for light and fast travel during the season when the soil is comparatively dry and firm. The objection to this method is that during the wet season the road-covering is injured by the large quantity of mud conveyed to it from the sides.

Opinions differ as to whether that portion of the carriage way to be finished and maintained as a dirt road, should be at the sides or in the middle. Heavy loads are apt to seek the sides, in order that the driver may walk upon the foot path, which favors metaling the wings rather than the middle.

It has been mentioned that the drainage of the road should be provided for by making it higher in the middle, and also by sloping it longitudinally. Engineers differ as to the most advantageous form of cross-section, some recommending a convex curve approaching to a segment of a circle, or a semi-ellipse, while others prefer two planes gently sloping toward the side-gutters, and meeting in the middle of the road by a short connecting convex surface. The latter method seems to carry the weight of testimony, the obvious objections to the convex road being that the water will stand in the middle of the road unless carried off by longitudinal slopes; that carriages will keep in or near the middle, and cause excessive wear along one line, in order to run on a level and avoid the tendency to overturn near the side-gutters; and that when travel is forced to take the sides, the labor of the horses and the wear of the wheels and of the road covering are greatly increased, in consequence of the oblique action of the weight, and the tendency of the vehicle to slide upon the road surface.

It is recommended therefore that the cross-section of the

road surface be formed of two straight inclined lines connected at the centre by an arc of a circle about five feet long, drawn to a radius of from 85 to 90 feet. The highest point of the arc should be in the middle of the carriage way. The degree of inclination toward the sides may be at the rate of 1 in 20 for rough roads, 1 in 30 for ordinary well-maintained gravel or broken stone roads, as in Fig. 18, and 1 in 40 or 50 for good paved roads. The drainage of the surface should, when practicable, be further facilitated by giving it an inclination longitudinally of not less than 1 in 125. In a level country this may be done at a trifling cost by a series of short gentle undulations.

## Catchwaters.

Upon a long stretch of continuously descending road, *catchwaters* should be placed at intervals. They are also necessary at the depressions where an ascending and descending grade meet, their province being to collect the water which runs down the surface of the road longitudinally, and convey it into the side-drains, thereby preventing the formation of furrows and gullies in the road surface. They are broad shallow paved ditches constructed across the road, and so formed that vehicles can pass over them without sustaining a severe shock. They may slope toward one of the side ditches only, or incline each way from the centre toward both, and, if located in a depression, will be placed at right angles to the line of the road. When placed upon a grade they should cross the roadway diagonally, in a straight line, when their discharge is on one side only, and if on both, their plan should be that of a broad letter V, with the angle pointing toward the ascent, so that they will arrest and divide the surface water and convey it to the two side ditches.

The catchwaters may have a descent of from 1 in 30 to 1 in 40, and, as their cross-section should be as nearly uniform as possible, the direction to be given to them in relation to the axis of the road will be governed by the steepness of the grade upon which they are placed, and the transverse form of the road surface. They should never be so located that one rear wheel and one forward wheel, on opposite sides of a vehicle, will enter them at the same time.

A mound of earth erected across the road, either in a straight line, or in a V shape pointing up the ascent, is a cheap substitute for a catchwater drain, and will answer very well if so proportioned that vehicles can pass it without inconvenience and with very little shock.

The pavement of catchwaters should extend to the point where the surface water is received—by the side-ditches or otherwise—to be conveyed away to the natural water-courses.

## Tools and Implements.

The most necessary *small tools* and *implements* used in the construction and repair of roads are hammers for breaking stone, forks for handling it, levels for adjusting the transverse form of the surface, and shovels and picks for general use.

The *stone hammers* are made of iron or steel, with wooden handles, and are of two sizes, one to be used sitting and the other in a standing posture. The first has a head 5½ to 6 inches long, weighing about 1 pound, fixed to a handle 18 inches long. The other hammer head weighs 2 pounds and may be 7 inches long, and has a handle about 3½ feet long. See Fig. 31.

The *fork* (Fig. 32) used in taking up the stones from the pile to load them into barrows or carts, or spread them upon the road, is made with ten or eleven stout steel prongs each 13 to 14 inches in length, set with their points 1¼ to

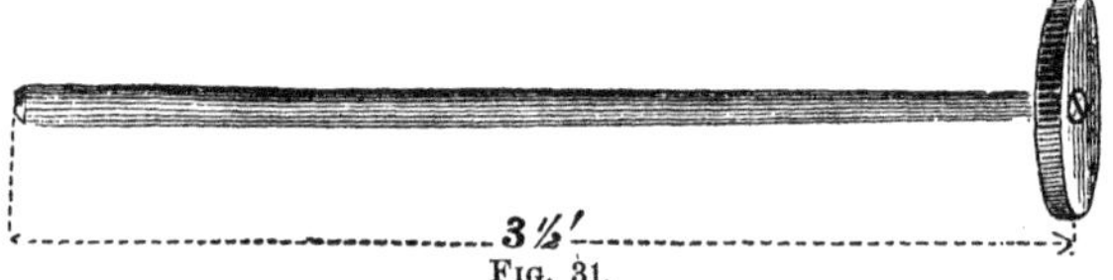

FIG. 31.

1½ inches apart. The whole length of the fork inclusive of handle should be about 4 feet 8 inches. Broken stone can be taken up with greater ease and rapidity with a fork than with any kind of shovel, leaving the detritus and earthy matter behind.

The *pick* is the one in common use, consisting of a bent

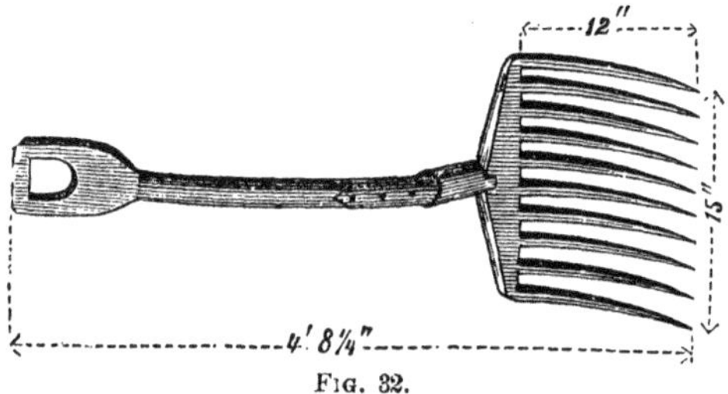

FIG. 32.

iron head tipped with steel at both ends and weighing about 10 pounds, set to an ash handle about 2¼ feet long. One tip in fashion like an adz, and the other into a blunt point.

The ordinary pointed *shovel*, with a concave blade and a bent wooden handle, is the most useful kind for road purposes.

The *level*, Fig. 33, is of great value in adjusting the cross-section of the road. It consists of a wooden straight edge A C with a plummet D at its centre. For a road 30 feet wide between the side gutters, the straight edge should be 16 feet long, which, omitting 6 inches at each end, may be divided into equal spaces—say five spaces of 3 feet each. Sliding gauges, set transversely in dove-tail grooves, are placed at the several points of division, omitting one at the end of the straight edge intended to rest directly on the centre of the road. By placing the level transversely across one half of the road, with the end, not occupied by a gauge, resting on the road surface at the crown, or at the precise height to which the crown is to be finished, and then mak-

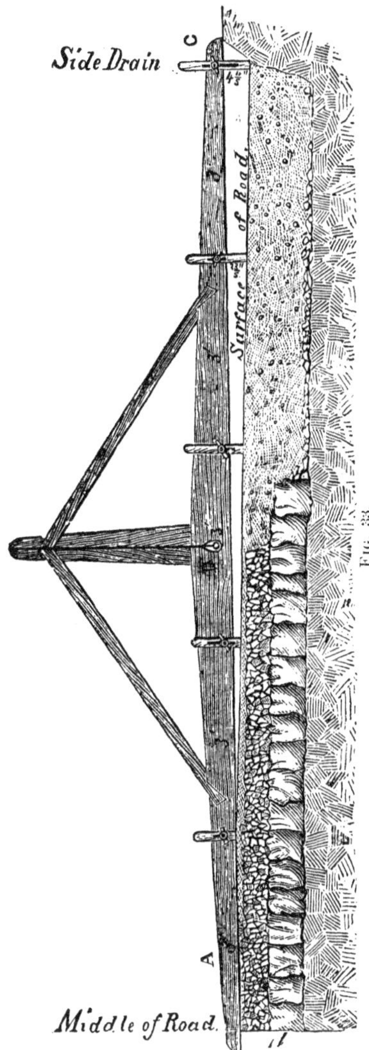

Fig. 33.

ing the straight edge horizontal by the plummet, the lower ends of the several gauges—previously adjusted and fixed to the required transverse slope—will indicate five points of the required road surface, the outer point being the bottom of the side gutter.

# CHAPTER III.

## ROAD COVERINGS.

ROAD COVERINGS, have for their object the reduction of the force of traction to the lowest practicable limit, at the least cost for construction and maintenance. They should be composed of hard, tough and durable materials, laid upon a firm bed, or upon an artificial foundation, from which water is excluded by suitable drainage.

The *basaltic*, the *doleritic* and other trap rocks, known also as green stone, the *sienitic* granites, and generally the hardest and toughest of the *feldspathic* rocks, and some of the *limestones* of the transition and carboniferous formations, furnish the most valuable road coverings, whether used in the form of rectangular blocks, or in small angular fragments, or as cobble-stones, gravel, or coarse sand. Slag from blast furnaces, and cinders and clinker from cement kilns, are also used for this purpose.

Wood in the form of blocks, or sawn into planks or slabs, is sometimes employed for the road-surface, but with little satisfaction or advantage. Its employment for foundations is sometimes expedient and even necessary in the absence of better materials.

The *covering* of most of the roads in the United States, and in all new countries, is the natural soil excavated from the side ditches and thrown into the middle of the roadway. In many cases, especially in sandy or gravelly soils, even the side ditches are omitted, and the road is simply a wagon track upon the natural surface, which soon becomes a broad shallow ditch, collecting and retaining the surface water from both sides of the track.

## Classification of Roads.

*Country roads*, as distinguished from *paved streets* in cities and towns, may be classified with respect to their coverings as follows :

1. Earth roads.
2. Corduroy roads.
3. Plank roads.
4. Gravel roads.
5. All broken stone (or Macadam) roads.
6. Stone sub-pavement, with top layers of broken stone (Telford).
7. Stone sub-pavement with top layers of broken stone and of gravel.
8. Stone sub-pavement with top layers of gravel.
9. Rubble-stone bottom with top layers of broken stone, gravel, or both.
10. Concrete sub-pavement with top layers of broken stone, gravel, or both.

## Earth Roads.

*Earth roads* necessarily possess so many defects of surface, that whatever amelioration their condition is susceptible of, by a careful attention to grade, surface-drainage, and sub-drainage, should be secured. The grades should be easy, not exceeding 1 in 30, the road surface should slope not less than 1 in 20 from the centre toward the sides, the side ditches should be deep and capacious with a fall of not less than 1 in 120, and trees should be removed from the borders to admit the wind and sun. In soils composed of a mixture of sand, gravel and clay, the road is formed of this material, and requires only that the ditches should be kept open and free, and that the ruts and hollows be filled up as fast as they form in the surface, in order to render the road a good one of its kind.

In loose sandy soils, a top layer six inches thick, of tough clay will be an effective method of improvement, which, to save expense, may be restricted to one-half the width of the roadway. Sand may be added to adhesive clay soils with equal benefit, the object in either case being to produce an inexpensive road-covering that will pack under the action of the traffic during the dry season, and will not work up into adhesive mud in rainy weather.

The material used in filling up ruts and hollows should be composed largely of gravel and coarse sand, free from sod, muck and mould. It should not contain cobble-stones, or larger fragments of rocks, which would form hard and unyielding lumps on one side of the wagon track, soon resulting in corresponding ruts and hollows on the other. All ruts should be filled in with good materials as soon as formed.

A pernicious custom prevails throughout a large portion of the United States, of repairing country roads only at certain seasons of the year. The cost of maintenance would be greatly reduced by frequent repairs, and especially by keeping the side-ditches clear and open to their full width and depth, by promptly filling in the ruts, and by maintaining the required slopes from the centre toward the sides. It will seldom be found that the material obtained by cleaning out the side-ditches is fit to put upon the roadway.

## Corduroy Roads.

Straight logs of timber either round or split, if cut to suitable lengths and laid down side by side across the roadway, scarcely deserve the name of road. They are nevertheless vastly superior to a soft marsh or swamp, which, in some seasons of the year, would be absolutely impassable for wheeled vehicles of any description. They are commonly known as *corduroy* roads from their ribbed character. In heavily timbered districts nearly all the logs for such a construction would be procured in clearing off the usual width of 4 rods—66 feet—prescribed for most country roads, the width of the road-covering itself, or the corduroy, being restricted to about 15 or 16 feet, so that two vehicles can pass each other upon it without interference. The logs are all cut to the same length, which should be that of the required width of the road, and in laying them down such care in selection should be exercised, as will give the smallest joints or openings between them. In order to reduce as much as possible the resistance to draught and the violence of the repeated shocks to which vehicles are subjected upon these roads, and also to render its surface practicable for draught animals, it is customary to level up between the logs with smaller pieces of the same length but split to a triangular cross-section. These are inserted with edges downward, in the open joints, so as to bring their top surfaces even with the upper sides of the large logs, or as nearly so as practicable. Upon the bed thus prepared a layer of brushwood is put, with a few inches in thickness of soil or turf on top to keep it in place. This completes the road. The logs are laid directly upon the natural surface of the soil, those of the same or nearly the same diameter being kept together, and the top covering of soil is excavated from side-ditches. Cross drains may usually be omitted in roads of this kind, as the openings between the logs, even when laid with the utmost care, will furnish more than ample water-way, for drainage from the ditch on the upper to that on the lower side of the road. When the passage of creeks of considerable volume is to be provided for, and in localities subjected to freshets, cross-drains or culverts are made wherever necessary by the omission of two or more logs, the openings being bridged with planks, split rails or poles laid transversely to the axis of the road, and resting on cross beams notched into the logs on either side.

## Plank Roads.

*Plank roads* were much in vogue twenty-five to thirty years ago, and are still used in localities where lumber is cheap, and stone and gravel scarce and expensive. They are usually about 8 feet wide, and occupy one side of an ordinary well-drained and properly graded earth road, the other side being used to turn out upon, and for travel during the dry season. The method of construction most commonly followed, is to lay down lengthwise of the road, two parallel rows of plank called *sleepers* or *stringers*, about 5 feet apart between centres, and upon these to lay cross-planks of 3 to 4 inches in thickness, and 8 feet long, so adjusted that their ends shall not be in a line but form short offsets at intervals of 2 to 3 feet, to prevent the formation of long ruts at the edges of the road, and aid vehicles in regaining the plank covering from the earth turn-out. New plank roads possess many advantages, for heavy haulage as well as for light travel, when the earth road is muddy and soft, but in a short time the planks become so worn and warped, and so many of them get displaced, that they are very disagreeable roads to travel upon. They are so deficient in durability that a common gravel road, as hereinafter described, will in the end be found more profitable in most localities. The ease and rapidity with which they can be constructed renders them a popular and even a desirable make-shift in newly-settled districts and towns where lumber can be procured at low cost, but they lack the essential features of permanence and durability which all important highways should possess. The sleepers ought always to be treated by some effective wood-preserving process to prevent decay. In the planks, ordinary rot will be anticipated by their destruction from wear and tear.

## Gravel Roads.

A capital distinction must be made between gravel that *will* pack under travel, and clean rounded gravel which will *not*, due to a small proportion of clayey or earthy matter contained in the former, which unites and binds the material together. Sea-side and river-side gravel, consisting almost entirely of water-worn and rounded pebbles of all sizes, which easily move and slide upon each other, is unsuitable for a road-covering, unless other materials be mixed with it, while pit gravel generally contains too much earthy matter. The gravel for the top layer at least, should be hard and tough, so that the wear will not pulverize it and convert it into dust and mud. It should be coarse, varying in size from half an inch to an inch and a half in largest dimensions. It should not be water-worn, and should contain enough sandy and clayey loam to bind it together firmly.

### Screening the Gravel.

Pit gravel usually contains so much earthy material that it should be screened, to render it entirely suitable for the surface layer. For this purpose two wire screens will be necessary, one with the wires $1\frac{1}{2}$ to $1\frac{3}{4}$ inches apart, while in the other they should not be more than $\frac{1}{2}$ to $\frac{3}{4}$ inch apart. The pebbles which do not pass the large screen are to be rejected, or if used, are broken up into smaller fragments; while the earth, small gravel, and sand that pass the smaller one, although unsuitable for the road surface, will answer for a sub-layer or bed for the road material to rest upon, or for side-walk coverings.

If the bed of the road is rock, a layer of earth should be interposed between it and the gravel, to prevent the too rapid wear of the latter.

### Applying the Covering.

In ordinary soils an excavation to the depth of 10 to 12 inches, and of the required width, is made for the reception of the gravel. The surface of this excavated form, called the *sub-grade,* may be made level, or preferably, it may be arranged parallel to the finished road surface by sloping it from the centre toward the sides. A layer 4 inches thick of good unscreened pit gravel in its natural state is first spread upon the road bed, which is then thrown open to travel until it becomes tolerably well consolidated. The gravel will usually be carried upon the road in wheel-barrows or carts, and adjusted to an even layer with rakes. The work may be hastened by using a cylindrical roller 2½ to 3 feet in diameter, and 5 to 6 feet long, weighing 1½ to 2 tons. A better design is to have two such cylinders arranged in a frame one behind the other, each being composed of two short cylinders 2½ to 3 feet in length, placed abreast upon the same axis. For compacting the bottom layers, and for the preliminary consolidation of the upper layers, a heavier roller cannot be used. A roller weighing from 5 to 7 tons and upwards may be used advantageously on the top layer, and if the light roller is not so constructed as to admit of loading, it would be well to have two.

The heavy roller constructed for the New York City Department of Parks, weighed six and a half tons, and could be loaded up to twelve tons. "It was composed of two hollow cylinders of cast iron, set abreast on a strong wrought-iron axle, making together a length of five feet, with a diameter of seven feet." The cylinders were set in a timber framework, and had apertures in the ends through which broken stone and gravel could be introduced into interior compartments, by means of which the aggregate weight could be increased to twelve tons. This roller is shown in Fig. 34, most of the shafts being omitted for want of space.

For making gravel roads a roller weighing 6 tons will suffice, the 12 ton roller being well adapted to the construction of broken stone or Macadam roads.

A horse road-roller designed and used in Germany is so arranged that the cylinder can be filled with water, when heavy rolling is to be done. When not in use, or when about to move from place to place the water is emptied out, and the weight materially lessened. A full description of this roller will be found in Mr. Clemens Herschel's "Treatise on the Science of Road-making," published with the executive reports of the Commonwealth of Massachusetts for the years 1869–'70. Road-rollers propelled by steam power have been used to some extent, but no description of them is deemed necessary here.

During the consolidation of the first layer by the light roller and by the traffic over it, men with rakes should be kept engaged in filling up the ruts as fast as they are formed.

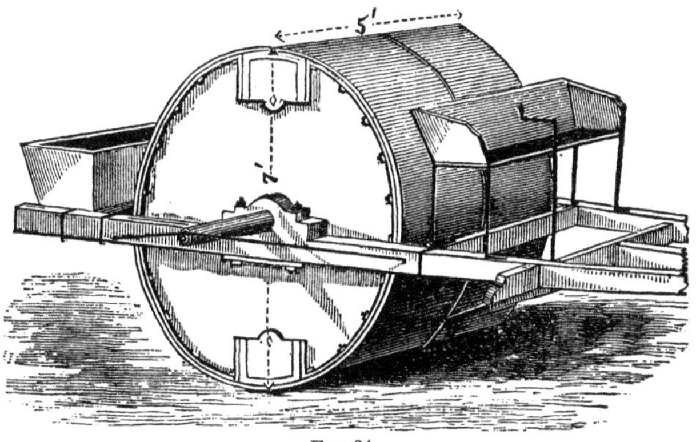


FIG. 34.

When the bottom layer is tolerably well, though not thoroughly compacted, a second layer of 3 to 4 inches is added and treated in the same manner. Successive layers follow until the road is made up to the required height and form of transverse section. The aggregate thickness of the several layers need not exceed 10 to 12 inches.

If the gravel be too dry to consolidate promptly it should be kept moist by sprinkling-carts, care being taken not to make it so wet that the earthy material will become semi-fluid and collect on the surface.

If the screened gravel of the top layer is so deficient in binding material that it will not pack firmly under the ordinary traffic on wheels, a thin layer, not exceeding one inch in depth, of sandy and gravelly loam and clay, or indurated sand and clay known as *hard pan*, or the scrapings from stone yards, should be spread over it and slightly moistened by sprinkling-carts before the rolling is begun, the light roller being invariably used first. A better practice would be to thoroughly mix the binding material with the gravel for the top layer before it is spread upon the road.

The sides of the road should be rolled first, to such degree of firmness that when the roller is placed upon the highest portion along the middle, the tendency of the gravel to spread and work off toward the side-gutters will be resisted. During the consolidation of the top layer the material must be kept properly moistened, and men with rakes should be in constant attendance to fill in ruts and depressions so as to give the surface the required form, and secure uniform density in the road-covering. When finished, the light coating of binding material will have been forced down several inches into the top layer, forming a kind of matrix which holds the gravel firmly in place, and provides a nearly watertight covering for the road bed. Gravel beds generally contain a greater or less quantity of large rounded pebbles, of 5 to 8 inches in longest dimension, which if not broken up into small fragments and incorporated with the road-covering, may be advantageously employed instead of sods, for facing the inner edge of the foot-path to protect it from the wash of the side gutters, and in forming the small drain under the foot-path from the gutter to the side-ditch.

## Good Gravel Roads.

A gravel road carefully constructed in the manner above described, upon soil of such sandy or gravelly character that the side-ditches will thoroughly drain the road bed to a depth at least 12 to 15 inches below the bottom of the road-covering, thereby rendering cross drains unnecessary, will possess all the essential requisites of a good road. In soils where the side-ditches will not secure good sub-drainage, cross-drains, which cost but little, should be introduced under the road covering at suitable intervals.

## Inferior Gravel Roads.

Country roads made with gravel are too frequently of a very inferior kind, being formed either by simply carting pit-gravel upon the road, and dumping it into the ruts and wheel tracks, and the gutter-like depressions worn by the tread of the animals, until the middle of the roadway gradually becomes covered to a width of 8 to 8½ feet, and a depth varying from 3 to 4 inches in the centre and 6 to 10 inches under the wheels, or by constructing an ordinary earth road with a single top layer of gravel 4 to 6 inches deep and 8 to 9 feet wide along the centre, the sides or wings being finished out with ordinary soil. Wheel ruts will form rapidly on such a road, which should be promptly filled with gravel. By this means the thickness of the gravel-covering will be gradually increased under the wheels to 8 and 10 inches and upward.

## Macadam Roads.

*Macadam* roads, Fig. 35, are constructed with successive layers of broken stone, applied in a manner similar to that above described for gravel roads. If the best quality of stone cannot be procured for the whole of the road covering, care should be taken to select the hardest and toughest stone for the upper, or preferably for the two upper layers, having an aggregate thickness of about 6 inches. The stone should be broken into fragments as nearly cubical in form as possible, the largest of which should not exceed $2\frac{1}{2}$ inches in longest diagonal dimensions. For inspecting the broken stone, an iron ring $2\frac{1}{2}$ inches in diameter may be used with advantage.

If the material be very tough and hard, like most of the basaltic and trap rocks and the sienitic granites, or if the traffic upon the road be light and its amount not large, the stone

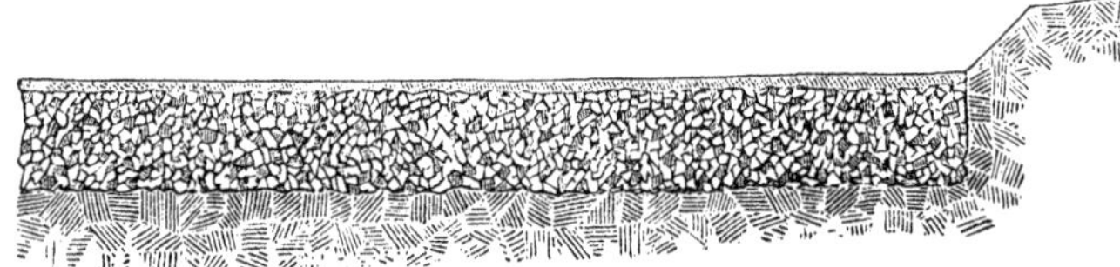

FIG. 35.

may be broken smaller, without danger of their crushing too easily or wearing too rapidly. The smaller the fragments, the less will be the volume of voids in the road covering liable to become filled with water and mud, and the sooner will the surface become hard and smooth when opened to traffic, or while being compacted with rollers.

In Macadam's matured practice upon the Bath and Bristol roads, England, he did not allow any stone above three ounces in weight (equal, with the material he had, to cubes of $1\frac{1}{2}$ inches or 2 inches in their longest diagonal length) to be used. He caused splinters and thin slices and spalls to be excluded as far as possible, and laid considerable stress upon uniformity of size, and perfect cleanliness or freedom from dust, sand, or earthy matter. The French engineers, on the contrary, are indifferent as to cleanliness, and upon their broken stone roads make use of all sizes from the dust and detritus produced in breaking, up to 1½ inch cubes, upon the assumption that the smaller particles occupy only a portion of the original void space between the larger fragments, the whole of which will sooner or later become entirely filled with dust and mud; an assertion which must be accepted in a modified sense, for it is certain that a well constructed and properly maintained Macadam road should allow, and will allow all surface water which finds its way through the crust of the road surface, to percolate freely through to the bottom of broken stone covering, where its prompt escape is provided for by suitable sub-drainage, upon soils where such a precaution is necessary.

The stone, if broken by hand hammers, will comprise fragments of all sizes, from the largest allowable dimensions, down to small particles and dust, and of various angular, prismoidal, and cubical forms.

It is not customary or necessary to screen hand-broken stone, but in loading it into the barrows or carts for transportation to the road, the detritus and most of the finer particles may be left behind by using the broad forks already described. (Fig. 32.) If the prevailing forms are angular, and of all sizes below the maximum prescribed, the fragments will unite and dovetail together more firmly and compactly than cubes, and very little if any binding material is necessary. If the smaller fragments and detritus be carefully excluded by screening, the road cannot be compacted into a smooth hard surface by rolling or by traffic.

Mr. Wm. H. Grant, Superintending Engineer of the New York Central Park, in his report upon the park roads, says, "At the commencement of the Macadam roads, the experiment was tried of rolling and compacting the stone by a strict adherence to Macadam's theory—that of carefully excluding all dirt and foreign material from the stones, and trusting to the action of the roller and the travel of teams to accomplish the work of consolidation. The bottom layer of stone was sufficiently compacted in this way to form and retain, under the action of the rollers, (after the compression had reached about its practical limit) an even and regular surface, but the top layer—with the use of the heavy roller loaded to its greatest capacity—it was found impracticable to solidify and reduce to such a surface as would prevent the stones from loosening and being displaced by the action of wagon wheels and horses' feet. No amount of rolling was sufficient to produce a thorough binding effect upon the stones, or to cause such a mechanical union and adjustment of their sides and angles together as to enable them mutually to assist each other in resisting displacement. The rolling was persisted in, with the roller adjusted to different weights up to the maximum load," (12 tons) "until it was apparent that the opposite effect from that intended was being produced. The stones became rounded by the excessive attrition they were subjected to, their more angular parts wearing away, and the weaker and smaller ones being crushed. The experiment was not pushed beyond this point. It was conclusively shown, that broken stones of the ordinary sizes, and of the very best quality for wear and durability, with the greatest care and attention to all the necessary conditions of rolling and compression, would not consolidate in the effectual manner required for the surface of a road, *while entirely isolated from, and independent of* other substances. The utmost efforts to compress and solidify them while in this condition, after a certain limit had been reached, were unavailing."

## Examination and Tests of the Stone.

In order to decide upon the fitness of any particular kind of stone for road covering, and especially when there are several kinds equally available, or so nearly so that the question of selection should be governed by the quality alone, an examination and tests of the varieties should be made, in order to determine their relative toughness, hardness, and power to resist abrasion. In some cases the difference of quality is so pronounced and so well known, that the opinion of intelligent stone-cutters of the neighborhood, who have been accustomed to work the several kinds of stone into various forms with different tools, will be sufficient to indicate their order, though not their degrees of merit, for the purposes in view. When, however, the information thus obtained is deemed inadequate or inconclusive, the examination may be continued as follows :

*First.* Average samples of the several stones—say a ton of each—should be collected together, and placed in separate piles in some convenient place. A practical stone-breaker should then be set to work with the ordinary stone hand-hammer, with directions to break up the material into sizes suitable for road metal, devoting an hour to each pile alternately, under the constant observation of the individual conducting the investigation. By this means the order of toughness of the several kinds of stone under examination will be ascertained, for it is the toughness, especially when in small fragments, which enables a stone to resist fracture from the repeated blows of a blunt tool harder than itself.

*Second.* The power to resist abrasion, or at least the order of quality in this respect of the several kinds of stone under trial, may be ascertained by grinding them under equal pressure, upon an ordinary grindstone run by power at a uniform speed. For this purpose bars of the same size, about 2–in. by 2–in. by 6–in. should be prepared from each sample under trial, in some of which the laminæ should be parallel to the end of the block, and in others perpendicular thereto. The blocks are then tested one after the other, by putting them endwise in a long box open at the lower end and closed at top, arranged vertically over the grindstone, with its lower end nearly touching the grinding surface. The box is firmly held in this position by framework attached to the grindstone frame. In the upper end of the box, directly above the specimen, there is placed a spiral spring, which is strongly compressed upwards against the top of the box, by the insertion of the stone specimen from the lower or open end. This spring supplies the force which presses the lower end of the specimen against the grinding surface, and of course exerts an equal force upon blocks of equal length. The amounts ground in equal times from the several blocks, as determined by their lessened weight, will give the inverse order of their powers to resist abrasion.

*Third.* The compressive strength of the stones in small (say 2–inch) cubes, although less directly indicative of their fitness for road covering than the foregoing tests, should also be ascertained, when it is convenient to do so, as corroborative data. With the information thus obtained it will not be difficult to make a judicious selection.

## Stone-Crusher.

Blake's stone-crusher, of which a longitudinal section and a perspective view of the essential parts is shown in Figs. 36 and 37, is an excellent machine for breaking stone for concrete or for road-coverings. AA is a frame of cast iron

FIG. 36.

in one piece, which supports the other parts. It consists of two parallel cheeks shaded dark in the drawing, connected together by the parts AA. B represents a fly wheel working on a shaft having its bearings at D, and formed into a crank between the bearings. It carries a pulley C, which receives a belt from a steam engine. F is a rod or pitman connecting the crank with the toggles GG. The end of the frame A, on the right of the figure, supports a fixed jaw H against which the stones are crushed. J is the movable jaw pivoted at K. L is a spring of India rubber which being compressed at each forward movement of the jaw J, aids its return. Every revolution of the crank causes the pitman F to rise and fall, and the movable jaw to advance a short distance toward the fixed jaw and return, so that a stone dropped in between the jaws J and H, will be broken

Fig. 37.

at the next succeeding bite. The fragments will then fall lower down and be broken again and again at each revolution until they pass out at the bottom. The bottom of the opening between the jaws may be set so as to deliver any required size of broken stone, by suitably adjusting the wedge N, inserted against the toggle block O. The crushed stone passes from the machine directly into a revolving cylindrical screen, inclined to the horizon, the meshes of which are small at the upper end, and of medium size, or about 2½ to 2¾ inches square in the middle and lower portion. The dust and small particles pass through the meshes in the upper end, while the large fragments which issue out of the lower end of the cylinder, are returned to the machine to be broken again. The rest is suitable for road-covering without any further preparation. The proper speed of these machines is about 200 revolutions of the crank per minute. They are made of several sizes, requiring engines of 4 to 12 horse power, and their working capacity varies correspondingly from 3 to 7 cubic yards of broken stone per hour. The best size for breaking road material is one having a capacity to receive stones 8 to 9 inches thick and 14 to 15 inches wide.

## Thickness of the Road-covering.

The thickness of the covering need not exceed 10 or 11 inches of well consolidated materials on a good road bed, for roads in cold climates subjected to the heaviest traffic. The French road engineers consider ten inches sufficient in France, upon the most important roads, and 6, 7, and 8 inches where the traffic is comparatively light. Macadam considered 10 inches of well compacted materials enough for very heavy traffic, and generally advocates less thickness than most English constructors. Six inches for the minimum and ten for the maximum thickness appear to have been his limits. In one instance he speaks of a road which "having been allowed to wear down to only three inches, this was found sufficient to prevent the water from penetrating, and thus to escape any injury from frost," and in another, states that "some new roads of six inches in depth were not at all affected by a very severe winter."

## Applying the Road-covering.

The drainage of the road bed having been provided for by side-ditches, and if necessary by suitable cross-drains, an excavation is then made to the sub-grade for the reception of the road materials, sloping from the middle toward the sides the same as the finished road surface, the depth of the excavation being regulated by the thickness adopted for the covering. It would be well, especially in made ground, to consolidate the bed by rollers, or by ramming.

A layer of broken stone three inches in thickness is then applied, care being taken, if dumped from carts or barrows, to spread it evenly with a rake. The road is then opened to travel in order that it may be compacted before the addition of more stone. This operation may be greatly hastened by rolling, beginning with the light and ending with the heavy roller. If the road bed be soft and yielding, whether naturally so at all times, or exceptionally so from recent rains, it may be necessary to omit using the heavy roller, for fear of forcing the bottom stone down into the soil.

Ruts must be carefully raked in as fast as they are formed. Experience has demonstrated that 3, or at most 4 inches of broken stone, is the greatest thickness that can be well compacted at one time.

## The "Wings" of Country Roads.

As it will seldom be necessary, except near large towns and cities, to apply the broken stone over a greater width than 16 feet, pit gravel, or sandy or gravelly earth may be used for extending the layer over the "wings.' This should be laid on and consolidated at the same time with the broken stone. When the lower layer shall have attained an even and tolerably well compacted surface, a second layer of stone not exceeding 3 inches in thickness, with gravel or earth on the wings, is then applied, and compacted by traffic and by rolling as before. The top layer is spread and consolidated in the same manner, but here the process of rolling should be prolonged, and an ample force of men should be kept constantly employed in filling in the ruts, and in removing lumps and elevations, so that the finished surface shall not only be hard and smooth, but accurately adjusted to the required gradients and transverse form. The roller should pass over every portion of the road surface from 40 to 60, or if necessary, even 100 times, and if the weather be dry the materials should be kept damp by sprinkling carts. A binding layer about 1 inch in thickness of gravel, or gravelly earth or hard pan, may be spread upon the top layer after it has become nearly consolidated, unless the character of the broken stone is such as to render this precaution unnecessary. When thoroughly consolidated, the finished road surface will not show any tendency to rise up and form a ridge in front of a 9 ton or 10 ton roller.

## Telford Roads.

These roads—named after Thomas Telford, by whom they were first constructed in Great Britain—are made with layers of broken stone resting upon a sub-pavement of stone blocks. Fig. 33 shows a transverse half-section of a road 30 feet wide, with a Telford covering 16 feet wide along the middle, and gravel wings.

Telford's specifications for a roadway 30 feet wide were as follows: "Upon a level bed prepared for the road materials a *bottom course or layer of stones* is to be set by hand in the form of a close, firm pavement. The stones set in the middle of the road are to be seven inches in depth; at nine feet from the centre, five inches; at twelve from the centre, four inches; and at fifteen feet, three inches. They are to be set on their *broadest edges and lengthwise across the road*, and the breadth of the upper edge is not to exceed four inches in any case. All the irregularities of the upper part of the said pavement are to be broken off by the hammer, and all the interstices are to be filled with stone chips, firmly wedged or packed by hand with a light hammer, so that when the whole pavement is finished, there shall be a convexity of four inches in the breadth of fifteen feet from the centre.

"The middle eighteen feet of pavement is to be coated with hard stones to the depth of six inches. Four of these six inches are to be first put on and worked in by carriages and horses, care being taken to rake in the ruts until the surface becomes firm and consolidated, after which the remaining two inches are to be put on. The whole of this stone is to be broken into pieces as nearly cubical as possible, so that the largest piece, in its longest dimensions, may pass through a ring of two inches and a half inside diameter.

"The paved spaces on each side of the eighteen middle feet, are to be coated with broken stones, or well cleansed strong gravel, up to the foot-path or other boundary of the road, so as to make the whole convexity of the road six inches from the centre to the sides of it. The whole of the materials are to be covered with a binding of an inch and a half in depth of good gravel, free from clay or earth."

## The Telford Sub-pavement.

For the sub-pavement the stone may be of inferior quality, as it is not subjected to severe wear and tear; but the toughest and hardest materials should be used for the top layer of broken stone.

The only advantage gained by setting the sub-pavement on a level bed, and gaining the required convexity of cross section by placing the deeper stones in the middle of the roadway, is a saving of expense in allowing the use of small stones at the sides. A better drainage of the road bed would doubtless be secured by making it parallel to the finished road surface, as was done with the Telford roads constructed in the New York Central Park.

The advantages and disadvantages of the sub-pavement or "bottoming," which forms the characteristic difference between the Telford and the Macadam roads, have been the subject of lengthy discussion between the advocates of these two methods of road construction.

It is alleged against the Macadam roads, that in compressible soils like clay, the weight of loaded wagons forces the stones into the earth; that in wet weather the clay rises up into the voids between the stone fragments, and prevents a thorough consolidation of the road covering; that in high latitudes the extreme cold of winter breaks up the road; that after a thaw the surface is liable to be cut up into deep furrows by the wheels; that during a drought the ordinary traffic upon the road causes a constant movement and consequently excessive wear among the broken stones; that there will also be considerable movement and therefore wear and tear due to the elasticity of the road bed, which cannot be entirely prevented by any ordinary thickness of broken stone alone ; and finally that the Telford bottoming constitutes a thorough underdrain, and besides being a remedy for all these imputed defects, is less costly than its equivalent of broken stone, as substituted by Macadam.

On the other hand, it is claimed by the partisans of the Macadam system, that the evils complained of do not exist to the extent alleged ; that suitable drainage will prevent them entirely ; that between the loaded vehicles above and the stone pavement below, the broken stone wears away much more rapidly than if laid directly on the earth ; and that generally a soft and elastic bottom is superior to a hard and unyielding one.

In constructing a road—whether a Telford or a Macadam —upon newly embanked earth, or any light soil that has not become thoroughly compacted, it is well to put the bottoming, or the lower course of broken stone, upon a layer of brushwood or fascines, in order that the settlement may be equalized as far as possible, and the formation of deep ruts prevented.

Rollers for compacting the road bed, before the bottoming is put down, and for consolidating the layers of broken stone, may of course be used in the same manner and with equal advantages upon Telford roads, or upon those where the covering is broken stone only, or gravel only.

## Telford Sub-pavement, with Gravel and broken Stone on top.

In some localities there may be an abundance of stone, such as sandstone and the softer varieties of limestone and gneiss, which is entirely suitable for the Telford bottoming but does not possess the requisite hardness and toughness for the top layer of broken stone. In such cases after the bottoming is set, the road may be finished with three to four inches of good gravel surmounted by a top or surface layer of good broken stone; or, the broken stone if too costly, may be omitted altogether, and the surface finished with a second layer of good gravel, in the manner described for gravel roads.

Whenever it is necessary to use an inferior quality of stone for the sub-pavement, the method of gaining the requisite transverse convexity by setting the smaller stones on the wings should not be followed, lest the road covering should fail there before it becomes seriously impaired in the middle. This precaution is specially applicable in cases where the amount of traffic is so great that the entire width of the road is used more or less constantly.

## Rubble-Stone Sub-foundation and Telford Pavement.

In soft ground it is very desirable that the foundation should possess sufficient firmness and unity of mass, to be able to resist any tendency to motion among the stones composing it, caused by the weight of passing vehicles, and the working up of the underlying soil into the interstices of the road covering. In order to secure this condition, a layer composed of rubble stones, varying in thickness from 3 to 5 inches, and in width and length from 8 to 18 inches, is sometimes placed upon the road bed as a foundation for the Telford sub-pavement. The stones are placed close together side by side flatwise, and rammed to their places, the interstices being afterwards filled in and leveled up with chips and spalls. A thin layer of sand or gravel is then spread over the surface, and compacted by ramming or rolling. Upon the foundation thus prepared, the Telford pavement is set, and the road is then finished with broken stone or gravel in layers, after the manner already described.

## Rubble-Stone Foundation without the Telford Pavement.

When the foundation is of rubble stones only, Fig. 38, it should, if the material is not too costly, have a depth of not less than one-half nor more than two-thirds of the entire

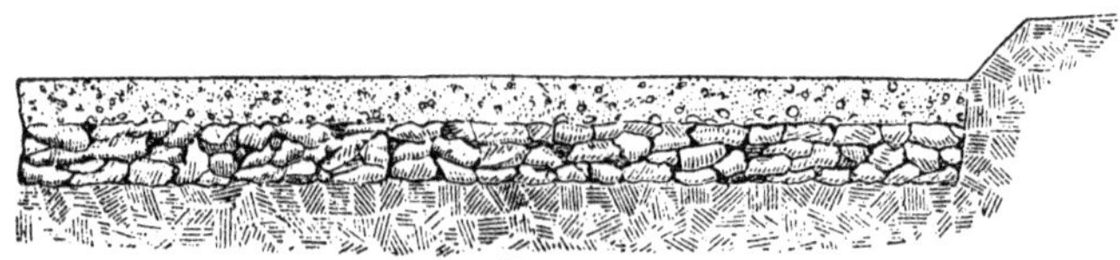

FIG. 38.

thickness of the road covering, whether the superstructure be of broken stone or gravel. For a total thickness of 10 inches of road covering, the rubble foundation may be from 6 to 7 inches thick, while 7 to 8 inches of rubble will not be too much for a road 12 inches thick.

The foundation should be constructed with great care, the larger stones being laid down first, side by side, flatwise upon the road bed, and firmly set to their places with rammers. The interstices are then filled in and leveled up with smaller stones, care being taken by selecting the pieces, to get them to fit as closely together as possible, and thereby to mutually sustain each other in place. The object is to use as much material as possible in a given thickness, so as to reduce the volume of voids to a minimum.

In placing the superstructure the first layer, whether it

be broken stone or gravel, should not exceed 2 inches in thickness, and it should be thoroughly compacted by rollers and by traffic before another is applied, in order that it may penetrate and unite with the foundation, and become indeed a part of it, during the process of construction. Otherwise there will be a subsequent tendency to work down into the rubble work unequally, causing ruts and depressions in the road surface. Moreover, it is of great importance that the foundation itself should remain firm and intact, and that the least motion among its elementary parts should be avoided, lest the stones should, in process of time, work up to the surface and destroy the road.

Macadam mentions the case of a road on Breslington Common, England, in the construction of which flag stones were laid down over the entire road-bed, and the road covering laid upon them. Their constant motion, or the slight tilting up of one end whenever a heavily loaded vehicle passed over the other end, kept the surface in a loose and unsettled state. Eventually they were found canted up and standing on their edges, and it was necessary to reconstruct the road.

Any possible motion in the foundation should be scrupulously guarded against, as likely to prove fatal to the stability and durability of the road. Where there is any reason to apprehend trouble from this cause, and indeed when the closest supervision cannot be had over the work, it will be safer to set the stones on their edge as nearly as possible after Telford's method, even should they be greatly dissimilar in size and shape, for an opportunity is then afforded to wedge in between them with chips and spalls, so as to guard quite effectually against their subsequent displacement from the effects of moving loads. The stones may vary in thickness from

3 to 6 inches, in width or depth from 6 to 9 inches along the middle of the road, and in length from 8 to 18 inches, without rendering it difficult to form them systematically into a sub-pavement, greatly superior in firmness and stability to any mere rubble-work foundation. Even flat cobble stones can be used, mixed in with the irregular fragments. The plan of such a foundation is shown in Fig. 40, and a vertical section transversely to the line of the road in Fig. 39.

FIG. 39.

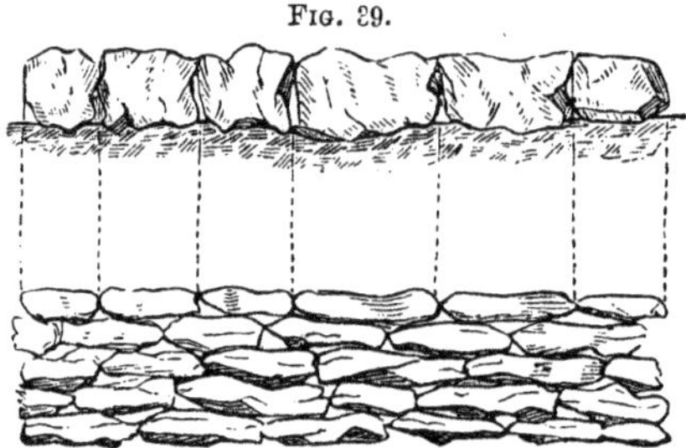

FIG. 40.

## Concrete Foundation surmounted with Gravel or broken Stone.

In soft, wet and elastic soils, liable to more or less constant saturation with water, and especially in cuttings through clay banks, and in other localities where the side slopes are badly infested with springs, the difficulty in the way of securing firmness and stability in the road foundation is frequently of very serious character, in consequence of imperfect sub-drainage. Parnell instances the case of the Highgate-Archway Road, near London, located between banks of clay where the soil was surcharged with water. Many fruitless attempts to drain the road bed had been made, a large quantity of broken stone had been used in the first instance, and subsequently taken up and relaid on gorse, brush and pieces of refuse tin. It was found impossible to consolidate the broken stone. It mixed up with the clay, and rapidly wore round and smooth, and the road finally became nearly impassable. It was rebuilt under the direction of Sir John MacNeill in the following manner: A thorough system of sub-drainage was applied by making four longitudinal drains throughout the entire length of the road, with cross drains at intervals of 90 feet. Smaller drains were placed 30 feet apart. On the road bed thus prepared to a width of 18 feet, a foundation of concrete 6 inches thick was laid. The surface of the concrete was indented transversely by a series of shallow triangular grooves, formed by embedding strips of wood in the concrete before it had set. These grooves were about 4 inches apart, and had a fall of 3 inches from the centre of the road to the sides, in order that any water which might percolate through the broken stone covering, would be promptly carried off. After the concrete had set, the superstructure consisting of six inches of broken stone was laid upon it, the wings or sides being carried out to the side gutters with flint gravel. By this means a dry and firm foundation was secured for the broken stone, and all possibility of any displacement of the latter by mixing in with the clay subsoil, or by the action of frost, prevented. The result, as might have been expected, was a first rate road. The concrete used for the foundation was composed of 1 part of Roman cement, and 1 part of sand mixed together dry. Eight parts of broken stone was then incorporated, using as little water as possible. From these proportions it is evident that there was not enough mortar in the concrete to fill the voids in the broken stone, while there doubtless was sufficient to bind the ballast together firmly, and resist the tendency to break up under the weight of loaded vehicles.

Concrete foundations, even if laid upon a level bed, should be finished on top with a slope from the centre to the sides, about the same as that given to the road surface, to facilitate the drainage of the top covering.

### Foundation of Rubble-stone and Concrete.

Reference has been made to the difficulty experienced in wet and elastic sub-soils in keeping a foundation of rubble stones firm and intact, and in preventing the stones working up and finally destroying the road surface. A remedy for this evil is found in the judicious use of hydraulic concrete between the stones, as shown in Fig. 41. In founding by

FIG. 41.

this method, the largest stones and those most nearly approaching the form of cubical and rectangular blocks, should be laid down first, side by side, but not in close contact, each stone being firmly set to its place by ramming. Concrete, in which the ballast should be composed of stone fragments not exceeding ¾ inch in longest dimension, or of a mixture of such fragments and pebbles of all sizes up to ¾ inch diameter, is well tamped in between and around the stones and carried up to the general line of their top surfaces. If a thickness of 6 to 8 inches is secured in this manner by one course of stones, this will suffice, and the road may be finished in the usual manner with layers of broken stone or gravel.

A foundation of this kind is believed to be as firm and durable as one of the same thickness composed entirely of concrete, while it costs considerably less. Its top surface should slope from the centre to the sides, in order to carry off all the water which percolates through the top layer of stone or gravel, a condition which can be secured either by sloping the road bed, or by selecting the larger or deeper stones for the middle and gradually decreasing their depth toward the sides, thus giving a greater thickness of foundation in the centre than at the sides.

It is of capital importance, in a foundation of this description, that the stones should be of such shapes that when set in place their side surfaces will be nearly vertical, or rather will be as nearly perpendicular to the road surface as possible, so that the concrete, after setting, will hold them firmly together, and effectually prevent any upward or downward movement, especially the latter, which might take place if the stones are of unsuitable shape or improperly set, as shown in Fig. 42.

FIG. 42.

If the stones very generally, or a great portion of them, are thin and slab-like in form, they should be set with their two largest and opposite surfaces cross-wise of the road and perpendicular to the road-surface, showing in vertical

FIG. 43.

longitudinal section as in Fig. 43. The concrete will then

hold them firmly in place, even upon a wet and spongy road bed.

## Shell Roads.

Upon the South Atlantic and the Gulf coasts of the United States, stone suitable for road coverings does not exist, and in most localities good coarse gravel or pebbles cannot be procured except at such an outlay for transportation as to practically exclude their employment for road construction. Oyster shells, however, can generally be had at from 4 to 5 cents per bushel, exclusive of land haulage, and when applied directly upon sandy soil, as a covering, 8 to 10 inches in thickness, they form an excellent road for pleasure driving or light traffic. They wear much more rapidly, of course, than broken stone or gravel of good quality, and require more constant supervision to keep them in good order. When properly maintained they possess most of the essential requisites of a good road.

## Charcoal Roads.

The novel expedient of using charcoal for road coverings is not likely to be resorted to except in newly settled, heavily wooded districts, where the standing timber has no market value, and must be gotten rid of before the land can be devoted to agricultural pursuits. A case is mentioned in Michigan where a good road was made through a swampy forest in the following manner:

"Timber from six to eighteen inches through is cut twenty-four feet long, and piled up lengthwise in the centre of the road, about five feet high, being nine feet wide at the bottom and two at top, and then covered with straw and earth in the manner of coal pits. The earth required to cover the pile, taken from either side, leaves two good sized ditches, and the timber though not split, is easily charred ; and when charred, the earth is removed to the side of the ditches, the coal raked down to a width of fifteen feet, leaving it two feet thick at the centre and one at the sides, and the road is completed." The material was found to pack well, not form into ruts, nor get soft and spongy in wet weather, although the water was not drained from the ditches. Its cost was $660 per mile, and contracts for two such roads were given out in Wisconsin at $499 and $520 per mile, respectively. (See Gillespie on Roads and Railroads.)

## CHAPTER IV.

### MAINTENANCE AND REPAIRS OF ROADS.

It is not considered to be fairly within the scope of this work, to enter upon a discussion of the methods by which the funds necessary for the proper maintenance of a public highway shall be raised and applied.

The turnpike system, however, under which those who make the longest trips are required to pay tolls for keeping up the road, is not believed to be equitable in all respects, nor the most advantageous to the community living on or adjacent to the line.

Many unthinking persons would be deterred from locating upon a turnpike, on account of the tolls to which they would be thereby subjected, regardless or ignorant of the fact that their haulage and other road expenses are likely to be greatly augmented by their unwise selection.

A judicious policy of road administration will attract population to the best roads, and therefore increase the amount of traffic to be accommodated, and correspondingly lessen the expense per capita for road maintenance. Any system which does not secure these substantial results, if not complicated by controlling circumstances of an adverse nature, must be either inherently bad, or inefficiently administered.

The advantage of maintaining a public highway in excellent condition, from motives of *economy* alone, is a question which rarely receives that careful attention from those having the matter in charge, to which its importance justly entitles it.

The average endurance or *life* of draught animals and of vehicles, are functions—calculable within reasonable limits—which enter directly into and should in a great measure control all considerations of policy on this subject, since they are not only not in conflict, but strictly coincident with the most advanced humanitarian views having a bearing on the question.

The traffic upon any given highway requires for its service a certain number of animals and vehicles, their number depending in great measure on the condition in which the road is maintained; and observation has shown that the amount of improvement in the surface of a metaled or other road, as ordinarily maintained throughout the United States, that would enable eight horses, for example, to do the work of ten without extra fatigue, is greatly below the estimate usually placed upon it by non-professional persons.

If, for instance, we take the case of a well-made broken stone road, clean and dry, and compare it with the same well-made road in a wet and muddy condition, we find that by Macneill's formula, page 29, a stage wagon weighing 1500 pounds, in order to carry a load of 1500 pounds at the rate of 5 feet per second (about $3\frac{4}{10}$ miles per hour), will require the constant exertion of a force of only $94\frac{3}{4}$ pounds upon the dry and clean road, while a force of $119\frac{3}{4}$ pounds will be required to move it at the same rate over the same road in a wet and muddy state. This increase of nearly 28 per cent in the force expended is due entirely to the fact that the road surface was not kept clean by sweeping off and removing the dust.

Hence if the amount of traffic on a given length of the clean and dry road required the daily service of 54 draught animals, their number would have to be increased to 69, to perform the same amount of service on the wet and muddy road. If the animals are driven singly there would be an addition of 15 drivers, and if in pairs one half that number. It would perhaps be fair to assume 5 pairs and 5 single horses, thus requiring 10 additional drivers on the inferior road. A yearly allowance of $225 for the purchase, feeding and care of each animal, and the purchase and keeping up of harness and vehicle, would probably be below the actual cost in those portions of the country provided, or which should be provided with metaled roads, amounting to $3,375 per year for the 15 extra horses and the equipments.

The hire and support of each driver may be set down at not less than $35 per month, or $4,200 per year for 10 drivers.

The aggregate amounts to the sum of $7,575 per year for extra cost of service upon a wet and muddy road, for the traffic of which 54 horses would suffice if the road were kept clean of dust, and consequently clear of mud.

During those seasons of the year when the inferior road is covered with dust only, but not with mud, Macneill's formula shows a difference of not quite 16 per cent in the force required to conduct the service, against the dusty as compared with the clean and dry road, equivalent, on the same basis of calculation used above, to an extra cost of about $4,300 per year for the service of animals, vehicles and men.

A draught animal, properly taxed, can accomplish upon a fair road 20 miles per day. from day to day, without unusual or excessive fatigue. If the road under discussion connects two towns 10 miles apart, one trip and return, carrying a load both ways, would be a day's task, the total amount of freight conveyed daily being the same whether the road be in good condition or otherwise.

It would be carried by 54 round trips daily on the dry and clean road, less than 63 round trips on the dusty road, and 69 trips during the seasons when the dust is converted into mud.

A fair average during the year, of the extra cost of service on the inferior road (amounting to the rate of $4,300 per year while the road is dusty, and to $7,575 per year when it is wet and muddy) will, of course, vary within certain limits, with the varying character of the seasons—with the wind, rain, sun, and temperature — but may, it is believed, be moderately set down at $5,000. The traction upon a well-constructed and well-kept metaled road, does not vary materially with varying moisture upon its surface.

We may therefore state the result of the foregoing discussion as follows: If the traffic between two towns connected by a well-maintained metaled road 10 miles long requires the constant service of 54 draught animals, the extra cost of conducting the same traffic will amount to at least $5,000 per year if the road be allowed to become covered and to remain covered with dust. This greatly understates the inevitable results of neglect, inasmuch as it assumes that the inferior road differs from the other only in the accumulation of dust upon its surface, while in point of fact it will soon wear into ruts and gutters which will convey the surface water into the road material, hastening the wear upon the surface, and greatly increasing the expense of haulage and the destruction of animals and vehicles.

The foregoing comparison has not been made between one road in a superior condition, having a hard and smooth surface, and another in a state so bad that it might be characterized as heavy, soft and rough; but the same well-conditioned road-covering has been under consideration; in one case kept clean, and in the other covered with dust or mud. A road of which the metal is in good condition generally, although covered with dust, is quite different from a rough, soft and heavy road, terms which imply that ruts, gullies, and inequalities of various kinds, all of which greatly increase the traction and the wear and tear upon animals and vehicles, have been allowed to form, and the dust and mud to accumulate upon the surface, a condition into which any good Macadamized road will degenerate in a very few years if neglected. If a road in this state be compared with one having a dry, hard and smooth surface, such as a well-maintained metaled road should possess, it will be found, whether the calculations are based upon the investigations of Sir John Macneill, or upon those of M. Morin, that an animal can draw about four times as much weight, vehicle included, over the good road, as he can over the bad one.

If, therefore, a suitable load for 2 horses over the good road be 4,000 pounds carried on a vehicle weighing 2,000 pounds,—total 6,000 pounds—only 1,500 pounds could be drawn by 2 horses over the bad road, rendering it necessary to add a third horse to draw the vehicle alone. These results are obtained in about the same ratio at all rates of speed not faster than an ordinary trot, and with all kinds of vehicles—carts, trucks, stage-coaches, and carriages for light driving.

If the traffic upon 10 miles of good road requires the

constant employment of 50 horses and 25 drivers, at an aggregate annual cost of $21,750, (putting the cost and support of men and animals the same as before,) it would cost $87,000 per annum, to conduct the same amount of traffic upon the same length of road covered with deep ruts and thick mud, and it would, beyond question, be a wise policy to expend the whole excess of $55,250, chargeable to the bad road, in improving and maintaining this road in a superior condition of smoothness and hardness, were such a large expenditure necessary to secure that result; for there would be saved thereby, not only this large amount, chargeable directly to extra men, animals, vehicles, etc., but money expenditures on other accounts not easy to estimate, together with the sacrifice or injury of local interests upon which it is difficult to put a money value; such as economy of time due to greater speed, a longer endurance for animals and vehicles; the advantage of lighter and cheaper vehicles; freedom from excessive dust and mud; and the increase of population, and therefore of traffic, attracted by better facilities for business intercourse.

## Relation of Animal Force to Traffic on Different Roads.

The cost of maintaining a road in good condition, under a given traffic, falls greatly below the extra cost of conducting the same traffic upon a bad road; the ratio between the two depending on local prices of labor and material, the quality of the road materials at command, and other circumstances not easily covered by a general rule.

If we assume that the amount of traffic between the two towns already referred to, requires the constant service of 50 horses with trucks weighing 2½ tons inclusive of load, upon a very dry and smooth broken stone road, then the additional horses required upon other kinds and conditions of roads, will be as shown in the following table, calculated from the results of M. Morin's experiments. The influence upon the force of draught exercised by the character of the vehicle, is omitted, as unnecessary in this discussion.

| KIND AND CONDITION OF ROAD. | Relative number of horses required to conduct a given traffic. |
|---|---|
| Broken stone road, very dry and smooth.......... | 50 horses. |
| Oaken platform, or plank road in good condition.. | 59 " |
| Broken stone road, moist and dusty.............. | 71 " |
| Causeway of earth, or dirt road in good condition.. | 93 " |
| Broken stone road, with ruts and mud............ | 112 " |
| Broken stone road, with deep ruts and thick mud. | 192 " |
| Solid causeway of earth, covered with gravel 1½ inches thick.......................... | 245 " |

It may therefore be adopted as a well established principle, that in all communities where the amount of traffic is sufficient to justify the construction of a good road of any description, or any road that is good of its kind, it should be maintained in a high degree of excellence, as a simple measure of economy.

## Macadam Roads seldom well kept up.

It is rare indeed, that a Macadamized or a gravel road is kept up in the thorough manner above indicated. In the great majority of cases the mud and dust are allowed to remain upon the road for long periods, and are seldom entirely removed; wheel ruts are allowed to form and enlarge, by reason of which not only is the resistance to draught, and the wear and tear of vehicles, greatly increased, but the surface drainage is destroyed to such extent that a large portion of the rain-fall collects in the depressions and finally percolates into the road covering; the side ditches become so obstructed that, in wet weather, water stands in them in places to the depth of a foot or a foot and a half, and upwards. The result is that during the wet season the road covering being at its foundation only a few inches, if at all, above the level of the water in the side ditches, and receiving by percolation from above a larger portion of the rain-fall, remains thoroughly soaked with water, which causes it to be soft and heavy. In this condition it yields readily to the wearing effects of traffic, loses its form on the surface, and soon becomes badly cut up with deep ruts and gullies.

The work of repairing a road in the condition above described, will be substantially the same, whether due to defective construction, subsequent neglect, or to both these causes combined; except where there was a failure to establish the necessary sub-drainage at the outset, in localities requiring it, in which case the repairs may practically amount to a re-construction of the road, or nearly so.

Assuming, therefore, that the road bed does not require to be disturbed, or, in other words, that it was suitably provided with cross-drains when constructed, or else from the character of the soil did not require them, the repairs should be conducted in the manner described in the pages which follow.

## The Maintenance of Broken Stone (Macadam) Roads.

The proper maintenance of a broken stone road consists in preserving the smoothness, hardness, and form of its surface, and thickness of covering, by a systematic restoration of the materials that are worn away by the traffic, and removed in the form of dust or mud.

The wear of materials is not in direct proportion to the average daily tonnage conveyed over the road, but increases much more rapidly than the tonnage, other conditions being the same.

## Two Methods of Maintenance.

Upon the roads in France, which have been the subject of prolonged and careful observation by the officers of the Corps des Ponts et Chaussées, two methods of maintenance are practiced, viz :

*First.* The method of minute daily repairs by which the road covering is preserved at a constant thickness ; applicable to roads of *moderate traffic* upon which the average daily tonnage does not exceed about 600 tons, upon a road covering 18 to 20 feet wide.

*Second.* The method of partial repairs, accompanied by periodical additions of material, by which the diminished thickness of road covering is restored at stated periods ; adapted to roads of *great traffic* upon which the daily tonnage exceeds 600 tons upon a road of the same width.

These two systems will be described separately.

## Maintenance of Broken Stone Roads of Moderate Traffic.

The thorough maintenance of a stone road of this class,

to such degree that extensive periodical repairs will not become necessary requires:

1. That it should be kept clear of dust and therefore clear of mud.

2. That thorough drainage should be maintained.

3. That minute repairs to the surface should be made systematically in small patches, as often as, and as soon as ruts or depressions begin to show themselves.

Under this method, properly followed, the thickness of the road covering will be maintained without diminution for an indefinite time.

The mud and dust, or dirt, should be cleaned from the surface and deposited beyond the side ditches, so as to expose the road metal slightly to view, without laying it bare, or removing the binding material from around the stones at the surface. This may be done by men suitably provided with hoes, stiff brooms set at right angles to the handles, shovels and wheelbarrows.

The hoes should have blades of hard wood, as those of iron or steel, unless used with the greatest care, might loosen up some portions of the stone, and needlessly and injuriously roughen up the surface. The brooms may be of birch, willow or other suitable wood.

The sweeping should not be so thorough as to remove the detritus, or binding material from around the stone slightly projecting above the general surface, so as to loosen them in their position, and endanger their being crushed separately piece by piece.

Draught animals instinctively follow in the track of preceding vehicles, the result being relatively excessive wear, and a tendency to form ruts along that track. Upon a road kept under watchful care this may easily be prevented in sweeping, by constantly effacing the wheel marks.

Machine-scrapers and brooms of various kinds, drawn by horses, have been used for cleaning the road surface, with considerable saving in both time and expense. It is necessary to use them with great care, in order to avoid loosening the stones at the surface.

Mr. Whitworth, of Manchester, invented a machine broom for sweeping up the mud and conveying it away. "It consists of a species of endless broom, passing around rollers attached to a mud cart, and so connected by cog wheels with the wheels of the cart that, when the latter is drawn forward, the broom is caused to revolve, and sweeps the mud from the surface of the road up an inclined plane into the cart." It is drawn by one horse and is said to clean the surface better, cheaper and more quickly, and with less injury to the road and less annoyance to passengers, than it can be done by machine-scrapers, or by hand labor.

In the city of New York, and other eastern cities, street sweepers of various devices have been employed, with greater or less saving of manual labor and expense. The one that has given the best satisfaction consists of a cylindrical brush or broom about 16 inches in diameter and 7 feet long, attached beneath the axle and connected by suitable gearing with the wheels of a two-wheeled vehicle drawn by one horse. The axis of the broom is set horizontally at an angle of about 40 degrees with the axle of the vehicle. The rear end of the broom is therefore about 4½ feet further from the horse than the front end. When working, the broom rests firmly on the surface of the pavement or road-covering and revolves in a direction opposite to that of the wheels, sweeping the dust and mud sidewise and leaving it in a ridge behind the rear end of the broom, thus sweeping a strip about 5½ feet wide. A second sweeper, or a second trip of the same sweeper if only one is used, moves the ridge of dirt 5½ feet further toward the side of the street and widens the part swept to about 11 feet. In this manner the dirt is finally delivered in the side gutters, where it is heaped up by hand with hoes,

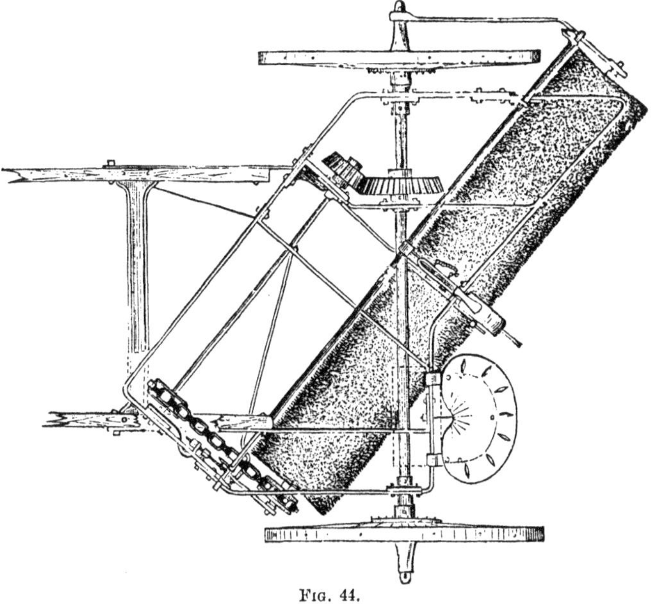

Fig. 44.

shoveled into carts and carried away. When at work the wheels and axle are rigidly connected, and revolve together. When not sweeping the broom is raised up a few inches from the ground and the axle is disengaged from the wheels, when both broom and axle cease to revolve.

This machine with 1 horse, 1 driver, and 10 men with

hoes, will do the work of 30 men with brooms and hoes, the shoveling into carts and carting away being of course the same in both cases. These data were obtained from an inspector of police engaged in the street-cleaning department in the city of New York, and are the results of prolonged and careful observation. A drawing of this sweeper is shown. Fig. 44.

In using this machine upon a road, the precaution should be taken to see that the brush is not too stiff. What would be entirely suitable, and in all respects well adapted for sweeping street pavements of stone blocks, wood, or asphalt, might injure the surface of an ordinary broken stone or gravel road, by penetrating too deeply, thereby loosening the stones at the surface and destroying the bond. It is important that the unity of surface should not be disturbed.

## Applying the New Materials.

The application of new materials to the road must be made not only with system and regularity, but under suitable precautions and restrictions, in order to combine efficiency and economy. Indeed the road should be always undergoing repair, in order that no necessity for making extensive repairs can occur.

"Every road should be divided into lengths, on each of which an intelligent laborer, who thoroughly understands his business, should be placed, to attend constantly and at all times to the proper state of the road, and for which he should be responsible. His office would consist in keeping the road always scraped clean and free from mud, in filling in any ruts or hollows the moment they appeared, with broken stone, which should be kept in depôts or recesses formed on the sides of the road, and one of which should be provided in each quarter of a mile. Those depôts should be capable of containing about 30 cubic yards of materials, and are best when the sides are formed with walls, so that the quantity of materials in them can be easily ascertained" (H. Law, C. E.).

Each of these men should be provided with a wheel-barrow, a shovel, a pickaxe, a scraper, a stiff broom and a rammer.

Upon roads that have lost their proper transverse form, a level of a length adapted to the width of the roadway, should also be provided.

During the autumn and spring, when the surface is soft and more work is necessary, additional men should be placed under the orders of the permanent laborers, but not in such manner as to divide the responsibility of the latter for the good condition of the road at all times.

The length of road to be given in charge of one man depends on circumstances varying greatly with the width of roadway, the character of the soil, quality of material used in repairs, etc. Three miles a man would not be too great a length upon some narrow country roads, one mile would be a full allowance upon others, while upon the great thoroughfares near large towns, a small fraction of a mile would be ample.

In France it has been found that one man can sweep in dry weather from 260 to 270 lineal yards of road, 5 to 6 yards wide, daily, if in a middling state, and twice that area if in an excellent state. If he had one and a half miles of road in charge, it could be swept from one to two times per month, according to its condition, which would be quite sufficient in most cases, thus leaving from 20 to 25 days in each month for other work, such as collecting and breaking stone, conveying it in wheel-barrows, to the road, spreading and compacting it, and keeping the gutters and side ditches free.

A machine-sweeper, if employed, could be used in common upon several of the sub-divisions of the road. As it would not thoroughly clean out the ruts and depressions, especially if these be of considerable depth, its work would, to some extent, have to be supplemented by hand sweeping.

The wear upon the surface of a well built road is slow, and so long as the vehicles can be prevented from following in each other's tracks, very even, a condition of things which can only be maintained by careful watching. A slight and apparently unimportant depression, if neglected, soon becomes a rut, in which the wear goes on with an increasing rapidity due to the increasing force of the blows imparted to it as it becomes deeper and deeper.

The new material should be added little by little, from time to time in the depressions and deficient places, and it should be broken fine, in comparison with that used in the original construction, containing all sizes and shapes upon to the largest, which ought not to exceed one inch and a half in longest diameter.

This method is strictly one of patching, and it should be done so constantly, that the small patches of broken stone will never exceed one to two inches in thickness, preferably not more than the thickness of one stone. If done when the road is firm and dry, the surface of the depressions to be filled should be loosened slightly with a light pick, to the depth of half an inch, so that the layer of new material may promptly become united with the old road, and some of the fine loose material can, with advantage, be taken out and spread over the broken stone as a binding. The loosening may be dispensed with in most cases, when the mending takes place soon after a rain, or after sprinkling, or when the road is in a soft condition.

Frequently the tendency to form a rut may be effectually arrested by sweeping into it the loose detritus from the adjacent parts of the road, and the free and expert use of the rake and broom will be found of great advantage at all stages of the work.

Penfold says that the ruts formed by wheels ought not to be filled up with loose broken stone, thus forming a ridge of materials possessing greater hardness than the parts immediately adjacent thereto, but that the rake should be worked back and forth across the rut and on either side of it, the object being to unite the old loose material with the new, in some degree at least, so that the patch will be as little unlike the unrepaired portions as possible. By this method of mending, a cubic yard of stone will usually suffice for a superficial rod of depressions and incipient ruts. The covering of large areas, exceeding eight to ten square yards, should not be undertaken at one time, and where there are several depressions in close proximity to each other, the worst should be patched first, and allowed to get even and solid, before the others are taken in hand.

" It is one of the greatest mistakes in road making that can be committed, to lay on thick coats of materials, and when understood, will no longer be resorted to. If there be substance enough already in the road, and which indeed should always be carefully kept up, it will never be right to put on more than a stone's thickness at a time. A cubic yard, nicely prepared and broken, as before described, to a rod superficial, will be quite enough for a coat, and, if accurately noticed, will be found to last as long as double the quantity put on unprepared and in thick layers. There is no grinding to pieces when so applied ; the angles are preserved, and the material is out of sight and incorporated in a very little time. Each stone becomes fixed directly, and keeps its place ; thereby escaping the wear and fretting which occur in the other case." (Penfold.)

Even in this patching process, rollers are sometimes employed for consolidating the stone, but by the judicious use of rammers weighing from 12 to 20 pounds, in conjunction with rakes and brooms, with which the wheel tracks are promptly effaced, and filled in, and the new stone slightly covered with detritus to bind it together, rollers may be dispensed with.

The deeper the depressions and ruts to be filled, the larger the fragments of stone used for repairs may be, up to the standard adopted for new roads. In a long continued season of drought, the road becomes baked and the metal begins to loosen after a while and consequently wears away with increased rapidity. In such cases great injury would ensue unless precautions are taken to fix the loose stone and restore firmness and stability to the surface layer. This can be done effectually by moderate sprinkling and light ramming or rolling, care being taken to so regulate the supply of water that it shall resemble a gentle shower of rain in its effect upon the surface, but not render the draught heavy. Too much water, by softening the binding layer on top, allows the stones to work upon each with increased grinding power.

This system of maintenance for roads of moderate traffic seems open to the objection of being unnecessarily expensive, but observation and experience have fully demonstrated that such is not the case, and that the "stitch in time" policy applies here with peculiar and significant force. It is not only vastly cheaper to maintain such a highway in good condition, for a given traffic adapted to it, than to pay the extra expense of conducting the same traffic on a bad road, but it is also vastly cheaper to keep the road in excellent order than it is to restore it to that state after a period of injurious neglect, during which it has become filled with deep ruts, and thickly covered with dust and mud.

A capital distinction must be made between the method here inculcated, which involves a constant and unceasing daily and hourly care of a road, in order to arrest every incipient tendency to deterioration upon its surface, and any and every other method whatever, whether by frequent repairs; or only occasional repairs; or by repairs at long intervals. The first only embraces the true principle, that of prevention. All the others are cures.

The French engineers of the Corps des Ponts et Chaussées were the first to give anything approaching to an exhaustive practical study to this question. It was found that in proportion as the intervals between the periods of repairs were shortened upon roads of small traffic, two important and valuable results invariably followed, viz., that the annual expense was lessened; and that the roads were always in better condition; and finally that the roads were never so good, nor the expense of maintenance so small, as when the system of unremitting and minute attention was in full operation.

Among the statistics bearing upon and elucidating this branch of the subject, room is here made for the following extract:

"The following took place with respect to the high roads of the Department de la Sarthe, somewhat less than 250 miles in extent.

| | | Per mile. |
|---|---|---|
| In 1793 a demand was made to put them in complete order for | £15,280, or | £60 |
| In 1824 the demand was above | 9,000, or | 36 |
| In 1836 the demand was above | 7,760, or | 31 |
| In 1839 the demand was above | 6,640, or | 26 |

And the roads have become better concurrently with the reduction in cost of maintenance, from being in 1793 in deep ruts, to 1839, when they were in very good order.

Part of the great road between Lyons and Toulouse, till 1833, was in a dreadful state, and yet it had cost habitually about £110 per annum per English mile for maintenance, when Mr. Berthault Ducreux introduced a system of patching instead of general repairs, since when, the road was gradually improved, till it was in a very good state, and the annual expense reduced, by £13 or £14 per mile" (Gen. Sir John F. Burgoyne, Bart.). During the period from 1833 to 1845 the traffic on this road never reached an average daily tonnage of 600 tons. From 1845 to 1852 it increased to over 900 tons; in 1856 it had reached 2500 tons with a steady increase still going on until it touched 2800 tons, when, upon the opening of a parallel railroad from Saint Etienne to Firminy, it fell off to about 2000 tons, which it maintained up to the last published reports in 1865. It was not therefore until 1845 and subsequent thereto, that the amount of traffic on this road was sufficiently great to render a change in the method of maintenance either necessary or expedient. This road will be further discussed when treating on the method of maintenance by periodical reconstruction with intermediate repairs.

In another case, that of the road from Tours to Caen it was reported in May 1836 to be in such bad condition as to require the expenditure of £2,000 on the entire line, to prevent the danger of its becoming impassable. It had been the subject of *occasional* repairs from 1832 to 1836, both included, at an annual cost of £978 for material and labor.

In January, 1837, M. L. Dumas of the Corps des Ponts et Chaussées was placed in charge of it, when in its worst condition. It was afterwards kept up by the method of constant and minute repairs. In August, 1838, it was reported, upon inspection, to be in a very good state, and it subsequently became better from year to year, at an average annual cost, from 1837 to 1841, a period of five years, of £820 for labor and material.

In 1834 the mail coach required five horses to draw it, and the service was so severe upon them that eleven died from over-work during that year. After 1838 only two horses were necessary to draw the coach, which they did without loss of animals or over-fatigue. Omitting all consideration of the cruel destruction of animal life inflicted by the bad road, there was in this case a saving, under the new and proper system of maintenance, of about 12 per cent per annum in the expense of labor and material put upon the road, and 250 per cent per annum in the amount of annual power required, for a specified item of traffic.

## Maintenance of Broken Stone Roads of large Traffic.

Experience upon the French roads seems to indicate that when the amount of traffic exceeds 600 tons per day over a road of 16 to 18 feet width of metal, the method of maintenance above described, by minute daily repairs, is not the most economical.

The wear of material is not in direct ratio to the tonnage passing over it. Other conditions remaining the same, it augments much more rapidly than the tonnage.

It is believed that for roads of great traffic the method of maintenance by periodical reconstruction, accompanied by partial repairs or patching during the intervals, offers superior advantages in respect to cost of labor and interference with traffic, than the one of minute and constant repairs.

This method consists in allowing the broken stone covering to wear down gradually and—so far as its wear can be controlled—evenly, limiting the work upon it to the preservation of its unity of surface, by filling in holes, depressions, and incipient ruts, with small quantities of stone, not rising above the general surface, and therefore not intended to restore that surface to its original height. The surface of the road is therefore constantly kept in good condition, and its resistance to draught at a minimum.

When the road covering, however, has been worn down to a thickness of 4 or 5 inches, a thorough repair to the extent of a restoration of the original thickness is made.

Experience has shown that the period of this general repair should not be deferred beyond a certain point, and that re-fillings of 3 to 4 inches in thickness are preferable to heavier ones.

The new layer, after being carefully spread to the required thickness, should be compacted with heavy rollers.

The season of least traffic is selected for this work, and, in order to lessen the interference with travel, the road-way may be covered and rolled upon only one-half its width at a time, and the rolling may be done at night.

This method has the advantage of providing a road surface which remains in a compact and regular shape for a long time after each periodical rolling is completed, during which only occasional and insignificant repairs are needed.

Some practical precautions, however, should be observed, viz.:

1. The re-laying and rolling should, if possible, be done in wet weather. If dry weather cannot be avoided, the road should be sprinkled rather copiously, so that the new material will unite readily with the old. A light picking up of the old crust will help materially to effect this union.

2. The new material after being carefully spread, should be repeatedly sprinkled and rolled until the stones are well pressed together into a compact layer, with an even and smooth surface. Then a layer of sand, or stone dust, or detritus, not exceeding half an inch in thickness, is spread over the top, lightly sprinkled with water and again rolled, in order to force it well into the voids of the new material. It is important not to put too much sand on at once. A repetition of thin layers is better than an excess in the first instance.

3. The wear of a road is represented by the decrease in the thickness of the stone crust, and the only reliable means

of finding this out correctly is to take occasional soundings to ascertain not only the thickness, but the composition of the road covering, with respect to solid material and detritus.

It has become customary to express the wear of a road in functions of its tonnage rather than of the number of collars, for the evident reason that a collar may represent only half a ton upon some roads and more than a ton upon others. But even the tonnage is not entirely reliable as a basis of comparison, and much depends on the kind of material used for the road covering.

The following table gives some of the results obtained on the roads in the Department of the Loire, France, from 1857 to 1860, as reported in 1865, by M. Graeff, engineer-in-chief des Ponts et Chaussées. The road covering was schist, and about 21 feet in width.

| Length of road. | Mode of Maintenance. | Daily Tonnage. | Annual wear in cubic yards. |
|---|---|---|---|
| 1 mile. | Periodical reconstruction....... | 1400 | 579 |
| 1 " | Minute and constant repairs.... | 1400 | 727 |
| 1 " | Periodical reconstruction....... | 1800 | 1866 |
| 1 " | Minute and constant repairs.... | 1800 | 2104 |
| 1 " | Periodical reconstruction....... | 2300 | 2794 |
| 1 " | Minute and constant repairs.... | 3200 | 4635 |
| 1 " | " " " .... | 5400 | 9934 |

Some of the roads in the arrondissement of St. Etienne, built with basalts, furnished the following data :

| Length of road. | Mode of Maintenance. | Daily Tonnage. | Annual wear in cubic yards. |
|---|---|---|---|
| 1 mile. | Periodical reconstruction....... | 1200 | 175 |
| 1 " | " " ..... | 2000 | 372 |
| 1 " | Minute and constant repairs.... | 2000 | 480 |

The foregoing tables show (1) that the destruction of road material increases much more rapidly than the tonnage; (2) that the tough basalts are much more valuable for road coverings than soft schist; and (3) that for roads of large traffic the system of maintenance by periodical reconstruction, accompanied by such intermediate repairs, more or less constantly, as will secure hardness and smoothness of surface, and uniformly diminishing thickness, is superior to the one of minute and constant repairs exclusively. Indeed, it is now generally admitted in France, that this last named system is not advantageously applicable to roads upon which the daily tonnage exceeds 600 tons.

## Curves of Tonnage and Wear.

The French engineers are very careful and methodical in collecting data, and arranging them in convenient form for practical use. Having ascertained by careful and systematic observation, the average daily tonnage, and the annual wear of materials upon different portions of a road of given width, or upon different roads, curves may be constructed with the average daily tonnage abscisses, and the quantity of material worn out annually upon any unit of length—say 1 mile—as ordinates. A separate curve for each kind of road material employed should be made, as shown in the diagram Fig. 45, which exhibits the results obtained on certain routes in the Department of the Loire, France, already referred to. The wear of four different kinds of stone was observed, viz., Schist, Quartz, Granite and Basalt. The width of the road covering was 21 feet. The lower horizontal line gives the average daily tonnage for different years, or upon different portions of the road for the same year, all measured from the zero point on the left. The left hand vertical line shows the yearly wear of road material in cubic yards per mile of road. The curve for schist was obtained by M. Graeff, who states that the deviation of the points D and E from the general curve may be ascribed to the mass of detritus brought upon the road by the exceptional traffic caused by the caving in of a neighboring tunnel on the St. Etienne and Lyons railroad. If the wear upon the road was proportional to the amount of traffic, the law would be expressed by straight lines instead of curves. The line for schist would pass through the points O and A, intersecting the ordinate corresponding to a daily tonnage of 5400 tons, at the point M, indicating an annual wear of only 2097 cubic yards, whereas the actual observed wear for that amount of tonnage was 9934 cubic yards. The observations on the wear of Quartz, Granite and Basalt were made by M. Montgolfier, and although not so numerous as might be desired, being limited to a daily traffic of 1800 tons for granite and quartz and 2000 tons for basalt, they still give an idea of the relative value of these materials for road covering. The ordinates of these three curves corresponding to the abscissa of 5400 tons daily traffic, were obtained by calculation, on the assumption that the ordinates of the several curves for any one abscissa bear the same rela-

10.000 cub. yds.
9000 c.y.
8.000 c.y.
7.000 c.y.
6000 c.y.
5.000 c.y.
4.000 c.y.
3.000 c.y.
2.000 c.y.
1000 c.y.
Annual wear of Schist per mile
Annual wear of Quartz per mile
Annual wear of Granite per mile
Annual wear of Basalt per mile
9934 c.y.
6.351 c.y.
5.083 c.y.
2.406 c.y.
4.637 c.y.
4.700 c.y.
2.654 c.y.
2.794 c.y.
1986 c.y.
1532 c.y.
1866 c.y.
900 c.y.
727 c.y.
G
H
I
K
M
Q
P
O
Tons
1200
1400
1500
1800
2000
2300
3200
5400

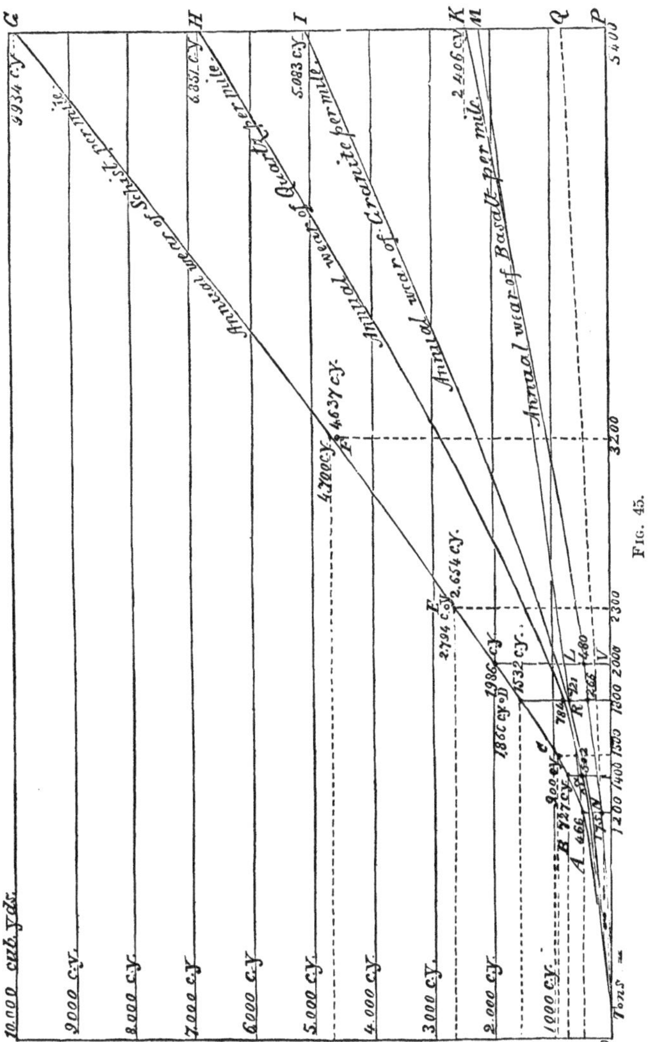

FIG. 45.

tion to each other as for any other abscissa. The following proportion gives the point K on the basalt curve.

480 : 1986 : : PK : 9934

PK=2405 cubic yards.

Besides the important deduction that no road can be maintained on the hypothesis that the wear is proportional to the tonnage passing over it, the diagram also shows that the necessity for employing only the best material, even when quite expensive, increases rapidly with the tonnage.

## Apportionment of Materials.

The apportionment of repairing materials, under the method of maintenance by periodical reconstruction, may be expressed by the formula $\mathbf{n} \times C = A + \mathbf{n} \times E$, in which $\mathbf{n}$=the number of years intervening between the periods of reconstruction or rolling.

C=the number of cubic yards of wear per year, per mile.

E=the mean number of cubic yards per year, per mile, for small repairs.

A= the mean number of cubic yards required for rolling. C and A become known for each road by careful observation for a sufficient length of time. E varies from year to year during a period of **n** years, being smallest in the year immediately succeeding a re-rolling, and largest in the year preceding the next rolling. In practice E will never exceed $\frac{1}{2}$ C, but will increase from zero to $\frac{1}{2}$ C in **n** years, with a mean average value of $\frac{1}{4}$ C, from which there results the practical rule of $\mathbf{n} = \frac{4}{3} \frac{A}{C}$.

It may be assumed that of the whole quantity of material

required for one period, by this system of maintenance, 25 per cent will be consumed in small repairs, and 75 per cent in one mass for re-rolling, and that the period lasts 2 years.

It is the opinion of M. Graeff that upon roads of large traffic this system of maintenance, as compared with the one of constant repairs, effects a saving of material that may, in some cases, amount to 40 per cent; and that during a period of three and a half years, upon certain observed roads in the Department of the Loire, the saving of labor was 10 per cent.

## Maintenance of Gravel Roads.

The maintenance and repairs of road-coverings composed of gravel, or of a mixture of gravel and broken stone, may be conducted upon the same principles, and by the same methods applicable to those of broken stone.

# CHAPTER V.

## STREETS AND STREET PAVEMENTS.

A STREET, in a city or town where the best ordered modern devices for promoting the comfort, convenience and health of the inhabitants have been introduced, should provide, upon and beneath its surface, (1) for the accommodation of ordinary travel and traffic, (2) for the drainage of the surface and subsoil, (3) for conveying away the fæcal and liquid refuse called sewage, and (4) for conducting water and gas to the inhabitants.

The subject of providing facilities for carrying on the necessary travel and traffic of a street, by suitably draining and paving its surface, is properly embraced in the design and scope of this work. A few brief suggestions with regard to sewerage and sub-drainage will not be out of place, before proceeding to a description of street pavements.

The importance of sewers in their relation to the health of a people cannot well be overstated. Those of ancient times were generally designed to receive and convey away both the fæcal refuse and surface water, and those of some of the best sewered cities of the present day have been planned and constructed with the same objects in view. The early sewers of England carried off the surface drainage only, the fæcal matter being generally collected in cess-pools located beneath or very near the habitations of the people, until the year 1847, when it was made obligatory to pass it into the sewers. In districts where the sewage is used for enriching the land, the question of its separation from the rain-fall may be an important one. On the other hand the surface drainage of streets that are closely built up, or where the traffic is heavy, is quite as impure, in time of moderate, or during the first stages of heavy rain-fall, as any sewage, and it might be unwise to allow all of it to flow directly into the fresh water courses of the neighborhood, in localities where the purity of those streams could be preserved by passing it into the sewers. In some cases the sewers take only the moderate rain-falls, the rain-fall pipes being so arranged that when their contents attain a certain velocity, the sewer ceases to intercept it in whole or in part. Circumstances of a local character, will control the plan and details of a system of sewerage, to such degree, that no universal rules for the guidance of the engineer can be laid down, although there are some general principles applicable to every locality. Even when the question of the manurial value of the sewage is to be considered, it will not always be judicious "to convey the rain-fall to the rivers and the sewage to the land," which is advocated by some sanitary writers as an unexceptionable rule.

Inasmuch as sewers are or should be water tight, as otherwise the contamination of the surrounding soil, and consequently of the atmosphere, by leakage, would be the certain result, they in no sense, when properly constructed, act as drains, by lowering the subsoil water-level. In well paved streets very little of the rain-fall is absorbed into the soil, but finds its way into the sewer, or other channels provided for it, and were it not for the unpaved areas, including back yards and unimproved lots, the question of draining the soil, in built-up streets would not, perhaps, possess great importance, especially if the soil be of a sandy or gravelly character.

It has been shown in Great Britain, from carefully prepared statistics, that the death rate from pulmonary diseases was reduced 50 per cent by sewering certain towns in such manner as to lower the subsoil water by drainage, while in other towns sewered with impervious pipes throughout, with no provision for drainage, there was no decrease in the death rate from consumption. Some provision for subterranean drainage should therefore be made without using the sewers for that purpose, although the laying of sewers alone, by cutting through the various impervious strata, invariably results in the drainage of the surrounding earth to a greater or less degree.

It is easy, when constructing the sewers, to arrange an effective system of subsoil drainage, generally at a moderate cost. There are several ways of doing this, among which the following may be mentioned :

*First.* The method by perforated inverts, gives when the invert blocks are laid, a series of continuous channels in the lower portion of the sewer. The joints between the invert blocks are left open on the sides and bottom, but are closely filled and pointed with mortar between the sewer and the longitudinal channels, to prevent the escape of sewage into the latter.

*Second.* Make the foundation of the sewer itself serve the purpose of a blind drain, by forming it of well compacted broken stone of various sizes. Between the broken stone and the earth on either side, a vertical layer of straw, hay, or fine brush may be placed, to prevent the choking of the drain with soil.

*Third.* Make a blind drain on each side of the sewer, by filling in with broken stone or a mixture of stone and coarse gravel, instead of ordinary soil.

*Fourth.* An ordinary drain of brick like any of those shown on page 56, or a tile drain, on each side of the sewer foundation will answer as well as any other, and can easily be laid at less cost than a blind drain of stone. Whatever method be adopted, it should be such as will secure a thorough drainage of the soil to the level of the floor of adjoining cellars. The areas in rear of the houses may be drained by either tile, brick, or blind drains, connected by a single pipe with the house drain, and thence with the sewer. As the arrangement of these drains, their relation to the soil pipes, and the location and design of the necessary traps and ventilating shafts, to prevent the escape of sewage gases into the houses, belong to the province of the sanitary engineer, no further reference to them need be made here. The treatment of the street surface will next be considered.

## Street Pavements.

It is desirable, for several reasons, that the surfaces of streets through large towns and cities should be paved.

The essential requisites of a good street pavement are, that it shall be smooth and hard in order to promote easy draft; that it shall give a firm and secure foothold for animals, and not become polished and slippery from use; that it shall be as noiseless and as free from mud and dust as possible; and that it shall be easily cleansed, and shall not absorb and retain the surface liquids, but facilitate their prompt discharge into the side gutter catch-basins. It should also be of such material and construction, that it can be readily taken up in places, and quickly and firmly relaid, so as to give easy access to water and gas pipes. Facility of repairs at all seasons of the year is another important requisite. Economy of maintenance and repair require that the material at the surface shall be durable.

All the road coverings heretofore described are wanting in one or more of the most important of these qualities, while they possess beyond doubt, some of those that are least essential, in even a greater degree than the best street coverings.

Road surfaces of broken stone or gravel, produce less noise, and give a more secure footing for horses than blocks of stone or wood, or a continuous surface of asphalt or other material, but they require such constant supervision to arrest the formation of ruts, and are so infested with either dust or mud, as to render them greatly inferior to a good stone or asphalt pavement for streets subjected to heavy traffic. An exception may perhaps be made in their favor upon suburban streets so extensively devoted to light travel or pleasure driving, as to justify the expense of frequent sprinkling by day and sweeping by night. These kinds of road coverings are also conceded to be excellent for the drives in public parks, and there are cases where the principal thoroughfare leading thereto should be constructed after the same method, and maintained with the same care, as the park drives, especially if the bulk of the travel and traffic over it be of a light character. They should be swept every night, and in dry weather sprinkled repeatedly during the day. Carts conveying the materials for repairs, whether gravel or broken stone, should be kept constantly upon the street, especially during those hours when it is least frequented, so that all ruts and depressions may be promptly filled as soon as they begin to appear.

It must be admitted, however, that there appears to be no trustworthy record of any urban street of this kind, in a thickly settled district, which has been maintained in such manner, that the inconvenience and annoyance inflicted by dust and mud upon the residents or people doing business on either side, did not in reality amount to a most serious public nuisance.

## Pavement Foundations.

The object of a pavement being to secure a hard, even and durable surface, and not to any considerable extent, nor necessarily, to support the weight of heavy loads, it is evident that the surface will soon subside unequally, forming ruts and depressions, unless it rests upon a firm and solid foundation. A good foundation is as necessary for the stability of a pavement as for that of any other construction.

Bad foundations invariably produce bad pavements sooner or later, while with a good foundation the quality of the surface upon which the wear takes place, depends upon the material used for paving, and the manner of laying it down.

Among the suitable foundations for a pavement, provided the thickness be adapted to the character of the subsoil and the nature of the traffic, are the following: (1) Hydraulic concrete, six to eight inches in thickness; (2) rubble stones set on edge, but not in contact, with the interstices filled in with concrete; (3) rubble stones set in contact, on edge, like the sub-pavement of a Telford road; (4) cobble stones firmly set in a form of sand or gravel; (5) small rubble stones of random sizes, in a well-compacted layer; or (6) a layer of broken stone laid in the manner of a Macadamized road.

Sand foundations are in most common use. They are described below, in the paragraphs on cobblestone pavements.

## Foundation of Broken Stone.

Broken stone foundations are prepared in all essential respects like a Macadamized road. They should generally be not less than 8 to 10 inches thick, or the usual thickness of a good road covering of broken stone. If the soil be of a yielding, soft nature like most clays, there should be a sub-foundation of sand or gravel, suitably rammed in layers, for the broken stone to rest upon. After the first layer of stone is spread upon the excavated road-bed or upon the sand form, the street may be opened to traffic, or, to hasten the operation of consolidation, rollers may be used upon it. A second and a third layer follow. In spreading the last layer, the required form of transverse section of the road surface is carefully established. The foundation is finished with a layer two to three inches thick of clean gravel, and the pavement is laid thereon, as hereinafter described.

## Foundation of Cobble Stones.

The cobble stone pavement set in a sand form, of which a description is given below, although furnishing a very inferior street surface for the reasons therein given, forms a good foundation for a pavement of stone blocks, and has frequently been utilized as such in the reconstruction of old cobble stone roads.

In setting the cobble stones less care would be necessary in their selection with a view to placing those of the same size together, than if they were themselves to form the road surface and sustain the traffic.

## The Cobble Stone Pavement.

The cobble stone pavement is the one in most general use in the United States, especially in new towns and cities, though entirely wanting in most of the essential requisites of a good street surface. It is formed of rounded or egg-shaped hard pebbles, varying in length from 6 to 10 inches, and in width from 3 to 6 inches. They are set side by side, in close contact with each other, with their smallest ends down, in a bed or form of clean damp sand or small gravel, previously compacted in layers upon the natural soil.

This sand foundation should be from 8 to 10 inches in depth, depending on the nature of the sub-soil. Before forming it, the road bed, after it has been excavated to the proper depth, is thoroughly consolidated by ramming or rolling. The sand should be compacted while in a moist state.

The cobble stones, after being set in position, are firmly settled to their beds by a heavy rammer, so as to bring their tops to the required roadway surface. Several rammings are sometimes necessary to secure their even adjustment. It is usual to give the required convexity to the surface—about 1 in 40 to 45 from the centre to the side gutters—by placing the largest stones in the middle, and suitably graduating the sizes toward the sides.

After the pavement is laid a layer of sand or fine gravel, two or three inches thick, is spread upon the surface and allowed to work its way in between the stones.

The defects of this kind of pavement are that its resistance to traction is great, while it is noisy, rough, and difficult to clean. The stones are liable to be pressed down unequally into the sand foundation, resulting in ruts and depressions which necessitate frequent repairs. It is severe upon vehicles and animals, and very unpleasant to travel over.

In laying cobble stone pavements in the city of New York, the usual requirements are that "The paving stones must be heavy and hard, and not less than six inches in depth, nor more than ten inches in any direction. Stones of similar size are to be placed together. They are to be bedded endwise in good clean gravel, twelve inches in depth. They shall all be set perpendicularly, and closely paved on their ends, and not be set on their sides or edges in any cases whatever."

Sand is unsuitable for a foundation, except when in a confined position where it cannot spread or escape laterally, as is usually the case when compacted in the excavated road bed. It should be clean, and if mixed with gravel, screened from all grains exceeding one-fifth of an inch in diameter, and compacted in the foundation while moist, by ramming or rolling it in layers not exceeding four inches in thickness. When all these conditions are imposed it forms a cheap and tolerably good foundation for a pavement.

Sand from the sea shore, or beach sand, from which all earthy matter has been washed, cannot be thoroughly compacted by ramming, on account of the entire absence of cohesion among the grains, which causes it to slide from under and loosen up around the rammer at each blow. Sand of this quality should be consolidated by rolling, or if rollers cannot be had, clayey or earthy matter should be mixed with it in such proportions as experience in each case may suggest.

## Old Roads as Foundations.

An old road, whether it be paved, Macadamized or graveled, will generally be found to furnish a good foundation for a new pavement, care being taken to bring its surface to an even state, and to the required form, by removing all large elevations and depressions. If the old covering be cobble stones, the interstices at the surface should be cleaned out and then filled with clean sand or small gravel, well compacted, or, better still, with hydraulic mortar, or with concrete of which the ballast contains no fragments or gravel exceeding half an inch in largest diameter.

## Rubble Stone Pavement.

The rubble stone pavement (Fig. 46), resembling the uncoursed portion of some ancient *opus incertum*, is formed with fragments of stone of various shapes and sizes, laid closely and compactly together, so as to form as even a surface as possible, but not in lines or courses. It is superior to the pavement of rounded pebbles, inasmuch as it may be made more even. It will therefore offer less resistance to traction, and be less severe upon vehicles and animals. The fragments of stone are such as can usually be selected, or produced with very little labor from the refuse of a stone quarry. The dimensions may vary from 3 to 6 inches in breadth and 6 to 12 inches in length, while the depth, to prevent their tilting up, should not be less than 5 or 6 inches. They are laid like cobble stones, in a form of sand or gravel, each stone being carefully adjusted to its place, so that when it

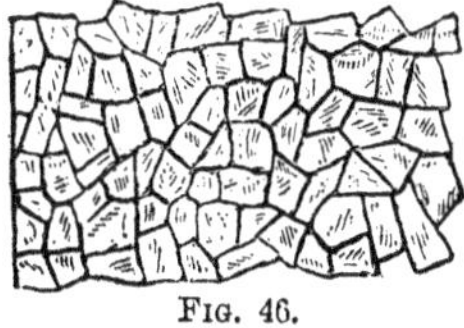
FIG. 46.

has been properly rammed its top face will coincide with the required surface of the pavement. Continuous joints in the direction of the draught should be avoided, in order to guard against the tendency to wear into ruts. To this end the long stones should not be set with their largest dimensions parallel to the sidewalks.

A rubble stone pavement laid in the manner above indicated, forms a good foundation for a pavement of stone blocks, and they may be laid upon a layer of sand or gravel about one inch thick, or in a bed of cement mortar, preferably the latter, although attended with some extra expense.

## Concrete Foundations.

Foundations of concrete, for street pavements, may be laid by the same method, and the concrete should be made after the same formula already laid down for roads, except that they should generally be somewhat thicker, to enable them to withstand the heavy traffic which passes over them in most cities and large towns. Upon firm and nearly incompressible soils, a thickness of 6 to 7 inches properly rammed in one or two layers, will ordinarily suffice, but in soils of a spongy, elastic nature, or largely composed of clay, a thickness of 8 to 9, or even 10 inches, will not be excessive. Though the most costly, it is the best street foundation, all things considered, that has yet been devised. In a few weeks after laying it becomes a strong, solid monolith, and even if it should crack in many places, in consequence of the great and varying loads upon it, or from unequal powers of resistance, and therefore unequal subsidence of the underlying soil, its superiority to any other kind of bottoming can scarcely be doubted. Perhaps not the least of its many advantages is the protection it affords against frost in high latitudes, subjected to long continued cold weather, such as prevailed in the northern portions of the United States, and in the Canadas during the winter of 1874–5. It was then observed that in Broadway, New York city, where the stone-block pavement rests upon a concrete foundation, the water and gas pipes were almost entirely exempt from injury by frost, while in the side streets, and notably in Fifth Avenue, which is covered with the Belgian blocks set in a sand foundation, the pavement had to be taken up to such an extent, and in such numerous places, as to cause serious annoyance to the traffic, to say nothing of the expense incurred in repairs, and the permanent injury to the street. It is impossible to take up a pavement in places, and relay it in the same condition in which it was found.

## Rubble Stone Foundations.

Rubble stone foundations for street pavements are constructed in essentially the same manner as for roads. Their thickness, however, should rarely be less than 8 to 10 inches, and they should not be resorted to, if the road bed is composed of easily compressible, or spongy soil, in which the stones comprising the lower layer would fail to find a firm and stable bed.

After the stones have been laid down to the required thickness, the surface should be made as even as possible, by breaking the stones of the top layer into small fragments, so as to fill on the surface interstices. For this work the long-handled hammer described on page 73, will be found to answer very well. There should then be spread on the surface a layer, 2 or 3 inches in thickness, of binding material, such as the detritus of the stone yards, or a mixture of clay, sand and gravel, or ordinary hard pan, or unscreened gravel.

The road may then be thrown open to traffic, or compacted by rollers, the ruts and depressions being constantly raked in. When the surface has become hard, smooth and even, a layer of about 2 inches in thickness of clear gravel is evenly spread thereon to receive the paving blocks. The top surface should be adjusted parallel to that of the finished street.

This foundation, therefore, is nothing more than a substantial road covering, consisting of a rather deep bottoming of rubble stone, surmounted with a thin surface finish of Macadam stone and binding material.

## Foundations of Rubble Stone and Concrete.

Pavement foundations of rubble stone, filled in with cement-concrete, formed after the general directions given on page 105, doubtless rank next to those of cement-concrete alone, in firmness and durability. Their thickness may vary from 6 to 8 inches, if the road bed be mostly clay of a yielding character, or if it be elastic or spongy.

In order to economize in the cost of the concrete filling, care should be taken when laying down the rubble stones, to adjust their upper edges somewhat evenly, so as generally to bring them into a surface parallel to that of the finished street.

## Sidewalks and Side Gutters.

For the convenience of foot-passengers, streets must be provided with sidewalks, on either side. Their width, which will depend upon the space that can be spared from the carriage way, the kind of traffic carried on in the locality, and the number of people requiring daily accommodation, should seldom be less than six feet, or more than fifteen. They are usually paved with flagging stones, brick, asphalt, wood, or cement-concrete, or some other variety of artificial stone, as described in Chapter VI. and should slope toward the street not less than 1 inch in nine or ten feet, in order that the surface water may be conveyed promptly into the side gutters.

The carriage way is separated from the sidewalk by a line of flagging-stones, sunk into the ground on their edges. These are called curb-stones, and form the outer side of the sidegutters and sustain the sidewalk, with the pavement of which their upper edges are set flush, so that the water can flow over them into the gutters. Their lower edge should extend at least 6, and preferably 8 or 10 inches below the surface of the street pavement, to which they act as a kind of abutment.

It is usual to pave with special care for a width of 14 to 16 inches, the lowest portions of the street on either side, called the side gutters, where the pavement meets the curb-stones. The slabs used for this purpose, called gutter-stones, or simply gutters, are laid flatwise, so that their upper faces form a part of the street surface. In the city of New York the curb-stones for the best paved streets are required to be not less than 3 feet long, 20 inches wide, and 5 inches thick; and the gutter-stones not less than 3 feet long, 14 inches wide and 6 inches thick. In streets crowded with traffic there is this objection to gutter-stones of uniform width, that the continuous longitudinal joint between the gutter and the rest of the pavement, wears into long, deep ruts, or grooves, which cause severe strains upon the running gear of vehicles, when the wheels, having once entered the rut, attempt to leave it. A remedy for this evil would seem to be to break the continuous joint, by making the gutter-stones of different widths—say 12 inches and 15 inches alternately, as shown in Fig. 47. Pavements for sidewalks are more fully described in Chapter VI.

## Pavement of Stone Blocks.

Although the form and dimensions of paving blocks have been the subject of much discussion, all authorities agree that the material should possess in a superior degree the qualities of toughness and hardness, so that it shall not crush, nor wear away too rapidly, under the effects of the traffic conducted upon it. It is also very desirable that the stones shall not polish and become slippery. While there may be considerable variation in the widths and lengths of blocks for the same pavement, they should be of uniform depth, or very nearly so, if the foundation be other than a form of sand or gravel. Were it otherwise the blocks of least depth would require to be underpinned at considerable extra cost, either with good mortar, or stone chips laid in mortar, in order to bring their tops to the required height. If a thick joint of sand alone be placed between them and the foundation, they will subside more than the blocks of proper depth which surround them, under which the layer of sand is very thin. The condition therefore that the blocks shall settle equally, requires them to be of uniform depth.

Hence, for pavements laid in mortar, the blocks should be of nearly uniform vertical thickness. As each stone is liable to have upon it the entire weight of the load carried by one wheel, it should be sufficiently large to resist crushing, and be so firmly supported underneath as to resist depression. In the direction of the draught, it should be no broader than the length of a horse's shoe, say not exceeding 4, or at most $4\frac{1}{2}$ inches, in order that the joints between the blocks may give a firm foothold at each step without slipping. Their depth in a vertical direction should be a little more than double their horizontal breadth, in order that they shall not tilt up on one side when a weight comes upon the other. For the same reason, and to increase the area of bearing surface on the foundation, their length across the street, should be at least equal to their depth, and may advantageously exceed it to some extent.

The most desirable dimensions for paving blocks are

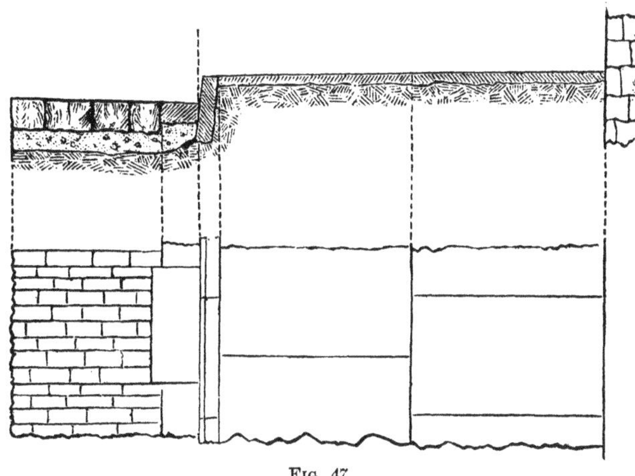

FIG. 47.

therefore as follows: $3\frac{1}{2}$ to $4\frac{1}{2}$ inches broad measured along the street; 9 to 12 or even 15 inches long measured across the street; 8 to 10 inches in vertical depth.

The stones are placed closely in contact, on their edge in

continuous courses, with their largest dimensions either directly across the street as in Fig. 47, or at an angle of 45° to 60° with its axis. It is claimed that the latter method is preferable, as the blocks are then less likely to wear into a convex form. When the joints run crosswise, the edges of the cross-joints receive a more severe impact from the wheels than when the latter cross the courses diagonally.

The stones of the same course must be of the same breadth, the broadest edge of each stone being placed down, when there is any difference in this respect. The joints are then close below and open on top, and should be compactly filled in with sand or fine gravel. Granite chips are sometimes wedged in between the blocks. Paving blocks generally give joints sufficiently open to secure a good foothold for horses, without rendering special care necessary to attain that end. When placed as in Fig. 47, the continuous joints run across the street and not lengthwise of it.

Fig. 48.

Upon steep streets the blocks are sometimes arranged in two sets of diagonal courses meeting in the middle of the roadway in an angle pointing up the ascent, as shown in Fig. 48. The joints, by sloping downwards to the right and left, aid the flow of surface water and other liquids into the side gutters.

If the foundation be concrete, or rubble stone filled in with concrete, or an old broken stone, cobble stone, or rubble stone covering, in good condition, it would be advantageous, and produce in many respects a better pavement, to set each stone firmly in a bed of stiff, though not very rich, cement mortar, care being taken not to disturb it again, or allow any travel upon it, until the mortar has had some days to set and harden. Some constructors recommend the use of cement mortar in the joints between the blocks. The mortar for this purpose should be mixed with a small quantity of water—not enough to make it plastic—and should then be tamped into the joints, or calked in after the method pursued in pointing first-class stone masonry.

The joints may be filled with bituminous mastic, or bituminous limestone, into which chips of granite or pieces of hard slate are compactly driven while it is warm and soft.

The usual method however, is to set the blocks in contact, and then to spread over the pavement a layer of clean sand, and allow it to work gradually into the joints. The sand however absorbs and retains the filthy surface liquids.

Sand being incompressible when in a confined state, constitutes a good filling, with the objection above named, especially when the blocks are set in mortar, as it then has no avenue of escape, and readily adjusts itself, and assumes a new condition of stable equilibrium, for every change or disturbance caused by the vibrations of the roadway.

When the foundation is a form of sand or gravel, or very uneven broken stone, or rubble stone, requiring a thick layer of sand to bring the surface to the required form, the paving blocks are not bedded in mortar, but are generally set in place upon a layer of clean sand or gravel spread upon the foundation to receive them. As the work progresses each block is slightly rammed, and when an area several yards square, reaching entirely across from one curb to the other has been completed, a heavy wooden rammer weighing 50 to 60 pounds manned by two men, is passed over them, a number of blows being given to each block. The blocks which break must be replaced by others, and those whose tops are forced below the required street surface, are taken up, additional sand filled under them, and then reset. A layer of clean sand about two inches deep is then spread over the pavement, and allowed to work its way gradually in between the blocks.

When the foundation is only a sand or gravel form, the paving stones should be somewhat larger in horizontal area than if intended for a concrete or broken stone foundation, for the reason that small blocks, and more especially thin blocks set on edge, have a tendency to settle into sand unequally.

Upon streets having a longitudinal inclination exceptionally great, special precautions may very properly be taken to secure a more perfect foothold for the horses' feet, than would be afforded by rectangular or cubical blocks placed horizontally with close joints.

If the blocks can be procured of marked wedge shape, without extra cost, it will generally suffice to set them with their broadest edge down, as in Fig. 49, so as to form a

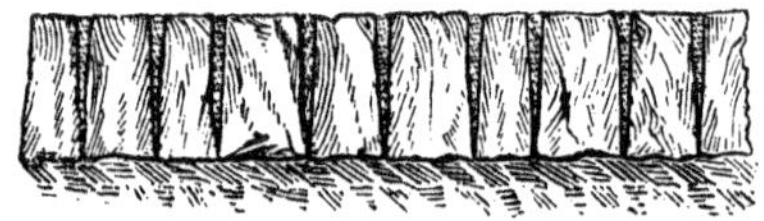

FIG. 49.

series of open joints across the street. These may be filled in, to within about one inch of the top, with granite chips firmly driven with a hammer, and topped off with a two-inch layer of clean sand. Some stone can be split readily into these forms.

With blocks that are essentially rectangular, the same end may be gained by setting the transverse courses about three-fourths of an inch apart, and interposing a course of slate between them as in Fig. 50, with the upper edges about

Fig. 50.

one inch below the street surface, finishing the work with a layer of sand as before, or, by a simple method, the stones may be set slightly canted on their beds so as to lean toward the descent as in Fig. 51, thus forming a series of triangular

Fig. 51.

ridges, or corrugations across the street. The joints are filled in with clean sand in the usual way.

## The Guidet Pavement.

Broadway, in New York city, below Fourteenth St., is covered with what is known as the Guidet pavement, composed of granite blocks as shown in Fig. 47, set on a foundation of cement concrete 6 inches thick. The same kind of pavement similarly laid, surrounds the New York post office, at the corner of Broadway and Park Row.

Upon newly made earth, or in wet, springy or swampy soils, the foundation should always be a layer of good concrete, at least 6 inches thick, laid upon a bed formed parallel to the finished street surface. The stones are then set in a layer of clean sand spread over the concrete to a depth of half an inch to an inch. The left hand portion of Fig.

47 shows this pavement. The usual specifications for the Guidet paving blocks require that they shall be of granite, equal in hardness to the Quincy granite, of durable and uniform quality, each measuring not less than 3½, nor more than 4½ inches in width, on the upper surface or face, and not less than 10 nor more than 15 inches in length, and not less than 8 nor more than 9 inches in depth. Blocks of 3½ inches in width on the face, to be not less than 3 inches in width at the base; all other blocks to measure on the base not more than 1 inch less in width or in length than on the face. The blocks are set upright in close contact on their edges, in courses, with the longest dimensions and the continuous joints running across the street, breaking joints lengthwise of the street. The ends of the blocks are dressed off so as to give close joints in the direction of the draught, while the broad vertical sides of the blocks are left rugged or uneven, or with the split rock-face, so that the continuous joints running across the street are somewhat open. This pavement, besides being firm, strong and durable, offers a good foothold for horses, in its open cross-joints, and an easy draught for loaded vehicles in the narrowness of the blocks. It gives very general satisfaction in New York, and seems well adapted to a street subjected to very heavy traffic. It would be an improvement, with a concrete foundation, to set the blocks in cement mortar, as a security against unequal settlement. It requires from 24 to 25 of these blocks to lay one square yard. Their cost, exclusive of land haulage from the dock to the street where they are to be used, varies with the price of labor, from 12 to 15 cents each, after allowing for loss and breakage. The cost of a sand foundation in large cities generally comprises excavating and rolling the

road bed, hauling away the excavated material, and purchasing, transporting, filling in, and rolling the sand, and will therefore, vary with the price of labor and sand, and the length of haulage. In the city of New York a sand form 9 to 12 inches thick, will cost, when ready for the pavement, from 40 to 60 cents per square yard.

A good cement concrete foundation 6 inches thick, exclusive of excavating and compacting the road bed and removing the materials, will cost from $1.40 to $1.50 per square yard.

At the present time, (autumn of 1875) contracts could be let in New York city, for the Guidet pavement on a sand foundation 6 to 8 inches thick, for from $4.75 to $5.25 per square yard. This includes a very liberal profit to the contractor.

In some localities, cubes of eight inches are preferred for paving. Their cost, delivered upon the streets in our eastern city, will generally not exceed $2.75 to $3.00 per square yard of surface. When laid upon a form of sand or gravel, 9 to 12 inches deep, it will cost at least 50 cents more per square yard to make the foundation and lay the stone, bringing the total cost of the finished pavement to $3.25 to $3.50 per square yard.

## The Russ Pavement.

Several years ago, a portion of Broadway, New York, was covered with the Russ pavement at a cost of $5.50 per square yard. The natural soil at the level of this road bed was sand slightly mixed with clay. This was excavated to a depth of 17 inches, and then a layer of granite chips, from 4 to 8 inches in largest dimensions, and about half that thickness, was laid upon the bed and rammed down nearly flush with the graded surface. Upon this was placed a foundation of concrete, six to seven inches thick, formed in detached rectangular sections, and composed of 1 volume of Rosendale cement, 2½ volumes of clean coarse sand, 2½ of broken stone like the Macadam road metal, and 2 of coarse gravel. The paving stones were rectangular blocks of sienitic granite, 10 inches deep, 10 to 18 long, and 5 to 12 wide, set in courses at an angle of 45° with the axis of the street, and so arranged that the pavement could be taken up in rectangular sections of 4ft. by 3½ft. in order to reach the gas and water pipes, without disturbing the adjacent portions.

This pavement did not give entire satisfaction, the surface of the blocks being too broad to give a good foothold for the horses. They also became smooth, polished, and slippery. It was therefore replaced by the Guidet pavement already described.*

* The following is condensed from the specifications in the patent granted to Mr. Russ, in 1846.

*First.* The sub-soil is graded.

*Second.* Granite chips, etc., 4 to 8 inches in diameter, and about half as thick, are laid on the road bed with the flattest sides upward, and rammed flush with the grading.

*Third.* A concrete foundation 8 to 10 inches thick is then laid in frames of sound wood, cast iron, iron stone pottery, burnt earth or any other fit material, thicker at bottom than at top. Before the concrete is put in, bars of iron are placed into the panels, crosswise, with holes in them, through which they are united by an eyebolt, with a ring in the head of each bolt. Large panels receive 2 or more sets of such bars, bolts and rings. The concrete is then filled in and consolidated, after which it may be lifted out of any panel to obtain access to sewers, gas or water pipes, etc.

*Fourth.* The pavement consists of granite or sienite blocks averaging 10 to 12 inches long, 4 to 5 inches wide on the top surface, and about 10 inches deep, carefully laid, the ranges of stones forming

## The Belgian Pavement.

The Belgian pavement, so named from its common use in Belgium, is made with blocks of stone that are nearly cubical in form, split as nearly as possible to square angles, with little if any dressing. The trap rock, which forms the palisades of the Hudson river, is extensively used for this purpose.

The following is one of the common forms of specification for this pavement; each block to measure on the face or upper surface not less than five inches nor more than seven inches in length; nor less than five inches nor more than six inches in breadth; in depth not less than six inches nor more than seven inches; nor shall the difference between the base and the top surface of any *stone* exceed one inch in either direction.

The sub-soil or other matter, other than clean sand, is to

lozenge-shaped divisions, presenting the edges diagonally to the wheel tires. The stone over the centre of each panel is to have 2 holes for a lewis, that it may be lifted out for a commencement of removing the stones to get access to the panel below. *This stone* should be set *only in clean sand;* all the rest of the stones are covered with sand that must be well washed into the joints, and there consolidated by thin grouting of hydraulic cement run freely into the sand and left to harden between the stones.

*Fifth.* I do claim as new and of my own invention the constructing a concrete foundation in panels or sections (to give access to pipes or conduits below) by the application and combination therewith of frames formed of any suitable material, with a thinner edge upward to allow the concrete mass to be lifted out when needed, substantially as described, when this is combined with a paved road-way of any kind laid thereon as described.

be excavated and removed to a depth of thirteen inches below the top surface of the new pavement when fully rammed, forming the proper arch or grade beneath the proposed pavement. Upon this foundation, clean, coarse sand or gravel is to be filled and thoroughly compacted by ramming or rolling to the proper depth to receive the paving stones, which are then to be laid in close contact, in even courses, transversely to the line of the street. When so laid, the pavement shall be thoroughly rammed to the grade and to the proper arch or crown, after which the surface is to be covered with one inch in depth of clean, coarse sand, and all interstices filled in solid with sand.

The blocks shall not be laid more than fifteen feet in advance of the rammers. When the road bed is not of firm and compact soil, the thickness of the sand foundation must be increased to 10 or 12 inches or more, and in soft, compressible, or swampy soils a concrete foundation should be resorted to. Indeed a solid and unyielding foundation is more necessary under a Belgian pavement, than under the larger blocks used in the Russ and Guidet methods.

The market price of Belgian paving blocks, whether of trap rock or granite, fluctuates with the price of labor, from $35 to $60 per thousand in New York city; so that the cost of the finished pavement laid upon six inches of concrete, will vary from about $3.60 to $4.50 per square yard, exclusive of profit to the contractor.

In setting the stones in a sand form it is important that they should all receive an equal amount of ramming, to prevent unequal settlement subsequently; and, if set in mortar on a concrete, rubble, or cobble foundation, they should not be walked upon or otherwise disturbed for some hours after they have been settled to their place, so that the mortar will have time to set, and the street should not be opened to traffic for some days, or until the mortar has attained sufficient strength to resist crushing.

## Wooden Pavements.

Wooden pavements made with blocks of wood—generally yellow or white pine—set on the end of the grain, although they have been extensively tried in the United States and elsewhere, within the last fifteen years, are unfit for streets subjected to heavy traffic. They are slippery in wet weather, and are of course very perishable, from their inability to resist either the wear and tear of traffic, or the course of ordinary decay. Various devices have been resorted to in order to lessen these objections and render these pavements safe and reasonably durable, such as setting them with wide open joints across the street so as to give the horses a good foothold; Kyanizing, Burnettizing or creosoting the wood to prevent decay; and underlaying them with an elastic foundation of boards or planks, to enable the blocks to resist the crushing and wearing effects of heavily loaded vehicles.*

* One of the most efficacious methods of preserving wood from decay, as well as from the attacks of land and marine insects, consists in impregnating it, by either the Bethell or the Seely process with the dead oil—containing carbolic acid ($C_{12}$ $H_6$ $O_2$) or cresylic acid ($C_{14}$ H $O_2$) obtained in the distillation of coal tar. "By the Bethell process the timber is placed in an air-tight cylinder of boiler iron, which is then exhausted of air to the point indicated by 20° on the Bourdon vacuum gauge. At this point the creosote is admitted into the cylinder at a temperature of about 120° Fah., at once filling it to within an inch or two of the top. A pressure of about 150 pounds per square inch is then applied, and maintained for from five to eight hours,

The usual sizes for the wooden blocks are from 3 to 4 inches in width, 8 to 14 inches in length across the street, and 6 to 8 inches in depth.

The ordinary requirements for the Nicolson pavement are that the block shall be of sound yellow or white pine, free from sap, of rectangular form, and not less than 3 nor more than 4 inches wide, not less than 6 nor more than 14 inches long, and 6 inches deep, the grain of the wood being in the direction of the depth. The blocks for paving the side gutters are to be sawed to a bevel, so as to form a suitable channel way of uniform cross section about six inches outside the line of curb-stones, to carry off the surface water.

In preparing the foundation, the subsoil and all material other than clean sand, is excavated to a depth of 9 inches below the top surface of the new pavement, and parallel thereto. Upon this road-bed clean sand is filled in to the required

depending on the size of the pieces of timber under treatment. The creosote oil is then drawn off and the timber removed."

The Seely process, in brief, consists (1) in subjecting the wood to a temperature above the boiling point of water, and below 300° Fah., while immersed in a bath of creosote oil, for a sufficient length of time to expel the moisture. When the water is thus expelled the pores contain only steam; and then (2) the hot oil is quickly replaced by a bath of cold oil, by means of which change the steam in the pores of the wood is condensed, and a vacuum formed into which the oil is forced by atmospheric pressure and capillary attraction. The dead oil referred to above contains only a small percentage of the two acids named. It is claimed that either of them applied in a pure unadulterated state to the surface of a piece of timber, like a paint, will thoroughly permeate the entire piece, even if it be one foot or more in thickness, and will effectually prevent decay, a question which has not yet been satisfactorily determined. The process is now on trial.

depth and a close flooring of common pine boards 1 inch thick is laid thereon, lengthwise with the line of the street, the ends resting on similar boards laid transversely from curb to curb. The flooring boards are thoroughly tarred on both sides with hot coal tar, brought to a proper consistency by boiling with pitch, so as to be tough and not brittle when cool.

Upon this flooring the blocks are set on end in parallel courses running across the street, the lower end of each block having been previously dipped to half its height in hot coal tar prepared as above directed. The joints which run parallel with the line of street are close, and not continuous.

The transverse courses are separated from each other $\frac{3}{4}$ of an inch, by batons of common pine one inch wide and $\frac{3}{4}$ of an inch thick, laid end to end at the base of the blocks, the whole being secured and made firm by nails driven through each baton and block with the flooring boards.

The spaces above the batons between the courses of blocks is then filled with a kind of concrete composed of clean roofing gravel and hot coal tar, thoroughly mixed and compactly rammed in with suitable iron-shod rammers.

Finally the surface of the pavement, as fast as it is finished, is thoroughly coated with hot coal tar, prepared as specified, and immediately covered with fine sand and gravel, mixed in equal proportions and laid on not less than one inch in thickness.

A section of this pavement taken parallel to the line of street is shown in Fig. 52.

A modification of this method consists in making the blocks square and all of the same dimensions on top (about

4 in. by 4 in.), one half of them being 3 or 4 inches less in depth than the rest, and then setting them on the prepared foundation in continuous courses closely in contact both across and lengthwise of the street, alternating the deep and shallow blocks with each other in both directions, thus forming a series of cells about 4 inches square and 3 to 4

Fig. 52.

inches deep, arrayed like the dark squares on a chess board. Fig. 53. These cells are filled with coarse gravel and prepared coal tar or asphaltum, and the whole pavement is coated over with the coal tar preparation and a layer of fine sand. When the flooring of boards is omitted, it is sometimes the practice to underlie the block with a coat of lime

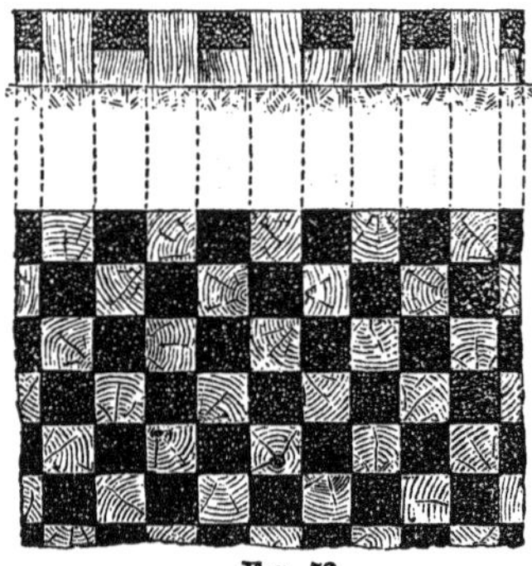

Fig. 53.

or cement mortar, or with a layer of thick paper covered with coal tar, in order to exclude the moisture from below.

By the Stowe method the blocks rest directly upon a form of clean well compacted sand or gravel, which may be from

4 to 6 inches in thickness only, if the road bed be hard and firm.

In soft spongy or clayey soils the sand foundation should be of greater thickness, though seldom exceeding 8 to 10 inches. The sand foundation should be carefully graded so as to be parallel to the finished street surface.

The blocks are set in courses transversely across the street

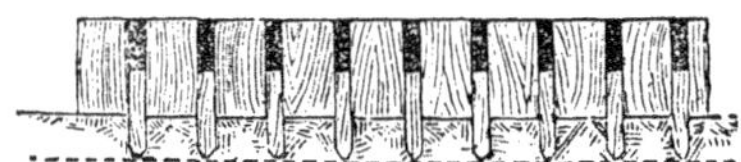

FIG. 54.

so as to break joints lengthwise of the street, the courses being separated from each other one inch, by a continuous course of wooden wedges placed close together edge to edge, and extending from curb to curb. These wedges are set in the first instance with their tops flush with the top surface of the blocks. After the whole pavement shall have been well rammed so as to give each block a firm bed, the wedges are driven down about 3 inches, and the open joints thus formed above them between the courses are filled in with a concrete composed of hot coal tar and fine roofing sand and gravel. (See Fig. 54.) The surface of the pavement may then be coated with coal tar prepared by boiling with pitch, and finished off with a thin layer of sand.

A modification of this general method of forming the pavement by setting single blocks of wood in courses slightly separated from each other, has been practiced to some extent without very satisfactory results. It consists in forming the blocks into sections or compound blocks, each containing from 12 to 15 single blocks breaking joints with each other in one direction, and held together with wooden tree-nails, passing entirely through the section from side to side, and striking each full block twice. Each section has a triangular groove running horizontally around its four sides, for the reception of strong wooden keys. The sections are set in contact, breaking joints, upon a foundation of sand or a flooring of boards, and thus mutually support each other and are prevented from tilting by the keys inserted horizontally between them.

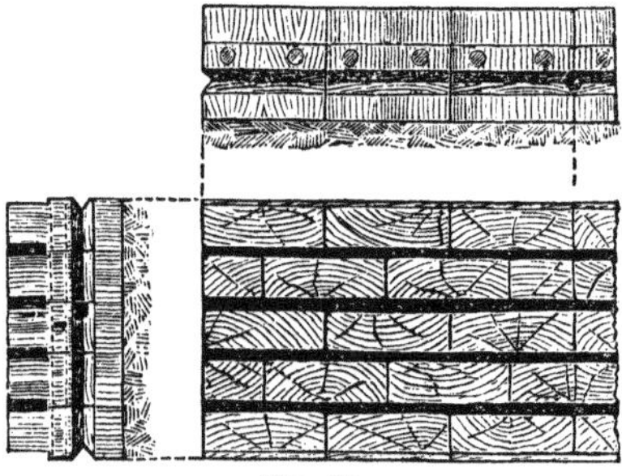

FIG. 55.

Deep grooves are cut in the upper face of each section so as to form continuous grooves across the finished pavement. These are filled with cement concrete, a mixture of coal-tar, sand and gravel, or other suitable material. Their principal object is to give a secure foothold for draught animals. One of the compound blocks is shown in Fig. 55. They are brought upon the street in readiness for lāying.

Another method is to set the blocks on a base of cast-iron formed into cells on top, into which the blocks, previously dipped in hot asphaltum, or otherwise properly prepared, are inserted to about half their depth, each block having a shoulder around it which bears upon the top of the iron partition separating the cells.

The cast-iron base is formed in sections, of convenient area for removal in places where necessary, each section being screwed to those adjacent to it by iron clamps or staples as seen in Fig. 56 at *a*, and by pins as seen at *b*; or they may be held together by dove-tail projections fitting into corresponding recesses. The blocks fit together so as to form close joints on top, a channel being cut in each, so as

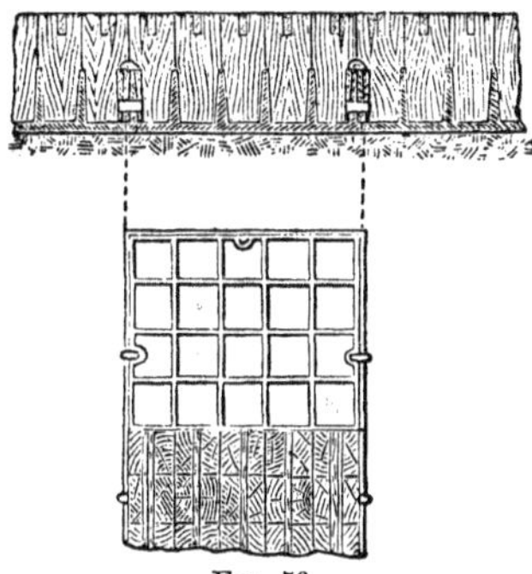

FIG. 56.

to form continuous channels, across the street, to give the horses a foothold. All the blocks are brought to the required form by machinery.*

* The author is indebted to E. S. Chesbrough, Esq., City Engineer of the city of Chicago, for the following interesting report on the wooden pavements in that city:

"It is true that wooden pavements have been more successful here than in any other important city. The reason is owing far more, I think, to local circumstances than to any peculiarity in the kinds of pavements used. Our city is on a very flat site, underlaid by a moist soil. Our streets are generally wide, and our traffic, instead of being principally on two or three main thoroughfares, is greatly diffused. When it is added we have the cheapest important lumber market in the country, you can easily see why wooden pavements should be more favored with us than they are elsewhere. Besides these con

## Brick Pavements.

Among the many attempts that have been made to combine bitumen and other hydro-carbons, including even ordinary coal tar, with other substances for street pavements, the hydro-carbonized brick pavement introduced by Messrs.

siderations I should mention that our wooden pavements, while in good condition, are much more agreeable to drive over, than any other kind it is practicable to put down here, except the asphalt, which is much more costly.

"As the property holders on each side of a street are assessed the cost of paving it, except at intersections with other streets, it is very natural they should have something to say with regard to the kind of pavement to be put down; in fact the City Council has of late allowed them to make private contracts, the work however to be subject to the supervision of the Board of Public Works. As a consequence of this state of things, and also of the rival claims of patentees, we have tried various kinds of pavement during the last twenty years; no kind has lasted any better than the Nicolson. There has not been time enough yet to determine whether some of the newer kinds will last as long or not.

"Of the kinds of pavement laid here the following includes all it would be of any practical value to describe, viz.:

"1st. The Nicolson, which you are familiar with.

"2d. Bastard Nicolson, differing from the real in the absence of wooden strips in the bottom of the joints between the blocks, which are filled with gravel.

"3d. The Stowe, which is without any flooring under the blocks. These are kept apart with thin pieces of wood, wedge-shaped at the end, and driven into the sand below. The upper parts of the joints between the blocks are filled like those of the Nicolson.

"4th. The Greeley, which differs from the Nicolson simply in the joint between the blocks, which is filled entirely with the wooden strips, leaving no room for gravel.

"5th. The unpatented, which is without flooring under the blocks,

Caduc and De Valins, of San Fancisco, California, deserves notice. The bricks are prepared by heating them in a boiler-iron tank set in brick work with a furnace underneath, and containing a sufficient quantity of the liquid hydro-carbon to allow for evaporation, and secure a thorough satura-

has no strips between them, and dispenses with tar. It is the most used here now.

"Some portions of wooden pavements here have lasted in tolerably good condition for nine years. When properly laid on wide streets they may be relied upon to keep in good condition about seven years, but some fail in less time, especially where there is much traffic and vehicles are compelled to go in ruts, as for instance, through the narrow arch-ways of our river tunnels, in which the wooden pavements keep in good order only two or three years. The traffic through these is not to be compared with that on Broadway, or Fulton St., New York.

"Gravel, Macadam, Cobble, and Limestone block pavements have all been used here, but none of them compare favorably with the best of similar kinds used in eastern cities. The cobble has been entirely discontinued. The Macadam is but little used. The stone is very disagreeable as compared with wood, and no new streets have been paved with it for years. The first wooden pavements of this city were laid in 1856.

"Our streets are under the care of the Superintendent of Public Works. Mr. I. K. Thompson, the former one, now a member of the Board, and Mr. Geo. W. Wilson, the present one, have kindly given me their views on this subject. They are both satisfied that the reason some wooden pavements last so much shorter time than others, is owing to a less degree of care or skill, in the selection of the blocks. This is of course, when they are subjected to like usage. They also think white oak is more durable than pine used in this way. The unpatented costs about $1.40 per square yard, including 4 inches of ballast blocks 8 inches in depth. The blocks are now laid diagonally across the streets, and are considered more durable laid this way."

tion of the bricks. The heat is applied with an intensity and duration sufficient to reduce the liquid material to a consistency that will withstand high atmospheric temperatures without softening. A tank 12ft. long, 4 to 5ft. wide, and 3ft. deep is a suitable size, and the time required for a thorough saturation of the bricks about 24 hours. It is claimed that, thus treated, the bricks will sustain great weight without crushing, and will resist abrasion and wear when subjected to ordinary street traffic, but no satisfactory evidence on these points, based upon their practical use, has been obtained. The description furnished by the inventors does not specify the kind of hydro-carbon required to give the best results.

The prepared bricks are set in a layer of sand about 2 inches thick, spread upon the road bed after the latter has been properly adjusted to a surface parallel with that of the finished pavement. The road bed and sand should be thoroughly compacted by ramming or rolling.

The bricks may be set in a single layer, on edge or on end, with continuous joints running across the street, and breaking joints in the other direction. They are firmly settled to their places by ramming, and hot coal-tar or asphaltum is poured into the joints to cement them together, and render the pavement impervious to water. The surface is finished with a coating of the same material laid on hot, and then top-dressed with a layer of fine gravel or coarse sand. When two layers are used, the bottom one may be of unprepared bricks laid flatwise, with the joints filled in with sand, and then covered with a coat of hot coal-tar or asphaltum. The top layer may be set on edge or on end, and finished off as in the case of a single layer.

On steep grade the surface layer may be set with open half-inch joints running across the street, filled in with a mixture of hot tar and gravel.

The durability of this pavement has yet to be proved.

## Asphalt Pavements.

Within the last twenty-five years bitumen, in some of its many forms, has been employed to a considerable extent, as the binding material or matrix for road and street coverings laid in continuous sheets without joints. They are all comprised under the general head of asphalt pavements. The city of Paris took the lead in this innovation upon the former methods of paving with stone, the reasons assigned for the change being, (1) the want of connection and homogeneity, in the elements of which the stone paving is composed, (2) the incessant noise produced by them, (3) the imperfect surface drainage which they secure, by reason of which the foul waters are not carried off but filter into the joints, and (4) the ease with which they can be displaced, and used for the construction of barricades, breastworks and rifle pits in time of civil war.

## Varieties of Bitumen.

There are several varieties of bitumen which pass insensibly into each other, from *naphtha* the most fluid, to *petroleum* and *mineral tar*, which are less so, thence to *maltha* which is more or less cohesive, and thence to *asphaltum* which is generally solid. The softer kinds gradually harden in time by the evaporation of the volatile parts.

They are hydro-carbons, accompanied in the solid and viscous kinds with various oxygenated carburets of hydrogen.

The fluid varieties are generally solvents, to a greater or less extent, of those that are solid or less fluid.

The forms of bitumen most extensively employed for pavements are *mineral tar ; asphalt rock*, which is an amorphous carbonate of lime impregnated with mineral tar, and known in commerce as *bituminous limestone ; asphaltum ;* heavy *petroleum oils* like those from West Virginia, or others not volatile under 212° Fah., or the residuum of refined petroleum containing no water, and so refined as not to be volatile at 212° Fah.

## Mineral Tar.

The principal sources of the natural mineral tar of commerce are in France, at Bastenne (Landes) and at Pyrimont Seyssel (Ain), and in Switzerland at Val de Travers, in the canton of Neuchatel. At Bastenne as well as at Gaujac, in the south of France, it flows from several springs mixed with water. The Bastenne mines are nearly exhausted, and no shipments of mineral tar to foreign ports are now made from that locality. This tar is also found impregnating quartzy sandstone, from which it is separated by boiling. At Seyssel both sandstone and limestone are impregnated with it. By boiling the sandstone in water the tar rises to the surface, or adheres to the sides of the vessel. The sandstone seldom yields more than 10 per cent of tar, the average of the mines falling considerably below that proportion. At the ordinary temperature mineral tar should not be either brittle or liquid, but viscous and ductile, so that it will freely elongate into threads when drawn out, and not break unless drawn very thin.

## Bituminous Limestone.

This limestone, known also as *asphalt rock*, occurs in both Seyssel and Val de Travers. The stone is of a liver brown color, irregular in fracture, massive, and has a specific gravity of 2.114, water being 1.000. It contains from 5 to 15 per cent, and sometimes 20 per cent, of the mineral tar above described ; is tough and difficult to break with a hammer, being to some extent malleable. It can be cut easily with a sharp knife, or scratched and abraded with the finger nail.

## Asphaltum.

Asphaltum is a variety of bitumen generally found in a solid state. At ordinary temperature it is brittle, and too hard to be impressed with the finger nail. It is black or brownish in color, opaque, slightly translucent at the edge of a new fracture, of smooth fracture, and has little odor unless rubbed or heated. It melts easily, burns with very little if any residue, and is very inflammable.

It is found floating on the Dead Sea, and in many places in Europe. Many localities in Mexico supply it, and it abounds in the islands of Barbadoes, Trinidad, and Cuba, and in Ritchie county, West Virginia, and in New Brunswick, Dominion of Canada. They all yield, by distillation, an inflammable gas ; a kind of bituminous oil ; a tarry substance like coal tar ; and a species of coke. They are all too brittle when cold, and too soft when exposed to the direct rays of a summer's sun, to be employed in their natural state, as the cementing substance of street pavements. A suitable solvent to render it fit for such use, has been found in the heavy petroleum oils, or the residuum of refined petroleum, not volatile at 212° Fah., already mentioned. The result of the combination, which should take place in an iron boiler at a temperature of about 470° Fah., is a manufactured bituminous cement exactly resembling in composition, quality, and appearance, the natural mineral tar obtained from the mines of Bastenne, of Gaujac, or from the sandstone of Seyssel.

This variety of bitumen, known as mineral tar, or bituminous or asphaltic cement, whether furnished by nature in a nearly pure state, or formed by the combination of other natural bitumens, under the general law that the more fluid kinds are solvents of those that are solid or less fluid, is the best and only suitable cementing substance, or binding medium, for asphalt pavements yet discovered. Like all the bitumens it owes its property of hardness and toughness under varying temperatures, to the presence in suitable proportions of the compounds called petroline and asphaltine, too much of the former making the asphaltic cement too soft in warm weather, while an excess of the latter renders it too brittle in winter. Hence the percentage of heavy petroleum oil necessary to be added in order to convert into a good bituminous cement, any particular variety of asphaltum, depends upon the proportion of petroline and asphaltine which the latter already contains, and this proportion varies greatly in different localities.

No asphaltic cement is suitable for all climates, and even the natural mineral tar from Seyssel, though well adapted for use upon the streets of Paris, requires to be mixed with a harder asphalt, to enable it to withstand exposure to the sun in the United States.

## Bituminous Limestone (Asphalt) Pavements.

A capital distinction must be made between pavements of asphalt hereafter described, made either with natural asphalt rock, or with the refined asphaltum as a cement, combined with suitable calcareous powder, and all or nearly all of those attempted imitations of it, produced by mixing crude mineral tar, or manufactured tar, with one or more pulverized minerals or earths. And more especially must we exclude from the category of asphalt pavements, all those patented street coverings composed of wood-tar, coal-tar, pitch, rosin, etc., mixed with either sand, gravel, ashes, scoria, sulphur, lime, etc., or with two or more or all of them. Some of them will produce a tolerably fair sidewalk, but they are totally unfit for the surface of a carriage way. Some of the best of them will answer for carriage way foundations.

The natural asphalt rock, like that from Seyssel or Val de Travers, or other localities in the Jurassic region, in order to be suitable for paving carriage ways, should contain about 11 or 12 per cent of bitumen (mineral tar) and 88 or 89 per cent of amorphous carbonate of lime. Inasmuch, however, as some of these limestones contain more, and others less than 11 or 12 per cent of the tar, it is frequently necessary and always practicable to obtain a mixture of the requisite degree of richness in bitumen, by combining those of different qualities together, or, if none but a rock poor in bitumen is procurable, the same result may be obtained by adding mineral tar to it.

The rock should be of the fine-grained variety, of tolerably close texture, and composed of pure carbonate of lime so uniformly and homogeneously impregnated with the bitumen, that a cut made with a sharp knife will show neither pure white nor jet black spots, but be of a brownish liver color, mottled with grey.

When asphalt rock of this character is heated to a temperature of 200° to 212° Fah., the bitumen becomes soft, the grains of limestone separate from each other, and the mass crumbles into a partially coherent powder. If this powder, while still hot, be powerfully compressed by ramming, tamping, or rolling, the molecules will again unite, and the mass when cold will assume all the essential qualities of the original rock, but in a superior degree, as regards toughness, hardness and incompressibility. This is the whole theory of asphalt road coverings, as applied to the street pavements in Paris and elsewhere.

## Foundations for Asphalt Pavements.

The pavement foundation should be preferably, cement, concrete, or rubble stones filled in with concrete made after the same formula, and laid in the same manner and to the same thickness, heretofore described for a pavement of stone blocks; or it may be six to eight inches of suitably compacted broken stone; or an old broken stone road carefully cleaned by scraping and sweeping, and then covered to an even surface with a coat of mortar; or an old cobble, or stone block pavement, with the joints raked out to a depth of about one inch, and then cleaned off and coated with mortar. Even a badly worn pavement of coal-tar concrete, or other kindred mixtures, may be the foundation, under suitable precautions. The mortar used for surfacing the foundation may be composed of one volume of common lime paste, one volume of the paste of Rosendale or other equivalent cement, and seven to eight volumes of coarse sharp sand. If standard Portland cement be used, the volume of lime paste may be doubled, and the volume of sand increased one-half, producing a mortar containing cement paste one, lime paste two, and sand ten to twelve.

If a new foundation has to be prepared, it should be of good cement concrete, and for streets subjected to heavy traffic not less than 6 inches thick if upon firm and compact soils, and rarely more than 9 or 10 inches if the road bed be wet, spongy, or clayey. It should be compacted to an even surface parallel to that of the new pavement, so that the latter can be applied in a sheet of uniform thickness.

## Heating the Asphalt.

The asphalt rock, previously pulverized by grinding, is brought to a uniform temperature of 250° to 260° Fah., in iron heaters, and in this condition is conveyed in wheel-barrows with sheet iron bodies to the place where it is to be used. If the heaters be arranged upon wheels so as to be portable, they can be kept in close proximity to the point of application, in which case the hot material may be taken out with long handled shovels, and deposited directly upon the foundation in its proper place, but it will generally be found convenient to convey it in wheel-barrows.

The material is not in fit condition for use until all the moisture has been driven from it, and it should not be applied upon a cement-concrete foundation until the latter has had some days to set, nor in any case upon any kind of foundation until its surface has become perfectly dry. If laid upon a damp surface the heat vaporizes the moisture.

and the steam, escaping through the powder, prevents a thorough and complete cohesion of the particles, and renders the pavement imperfect. When it is impossible or inconvenient to wait for the surface of the foundation to become dry, a first coating of asphalt, one-fourth of an inch thick, may be applied, to be followed when hard by the final covering. By this precaution the injurious effects of the steam created by the hot powder, will be avoided. For this lower coat the asphalt may be of poorer quality.

## Applying the Asphalt.

The hot powder having been carefully spread upon the foundation with an iron rake, to a depth exceeding by two-fifths the ultimate thickness required, is then compacted by

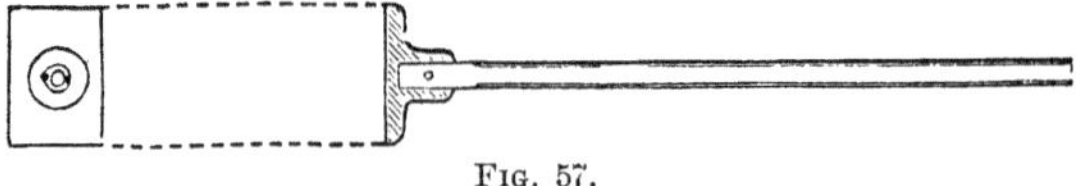

FIG. 57.

ramming with iron pestles, kept sufficiently hot in portable furnaces to prevent a too rapid cooling of the asphalt. The ramming should first be done cautiously with light blows,

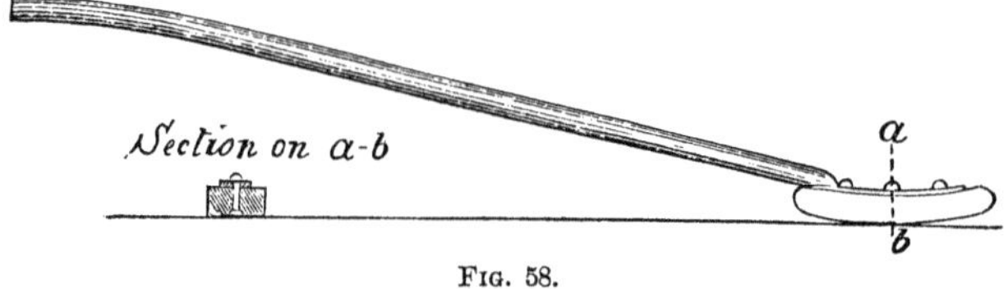


FIG. 58.

gradually increasing in energy, special care being taken at the junction of the hot powder with that previously laid. The finished pavement should be smooth and even, with from $1\frac{3}{4}$ to 2 inches thickness of asphalt, except upon streets where the traffic is very heavy, when the thickness may advantageously be from $2\frac{1}{2}$ to 3 inches. The iron rammer, Fig. 57, may be circular, square, or rectangular on the face, and should weigh from 14 to 16 pounds. After the rammers, hot smoothing irons, Fig. 58, are passed over the surface, in order to give it a high degree of finish.

It is usual to lay this pavement in transverse strips from curb to curb, or from gutter to gutter, of a uniform width of 4 to 6 feet, care being taken to cut the outer edge of each strip to a rebate while it is yet soft as shown in Fig. 59, so

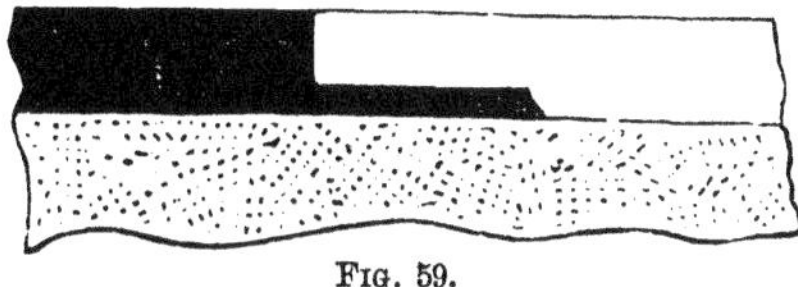

FIG. 59.

that the hot asphalt of the succeeding strip may lap over and firmly unite with it. If the strip has become cold, from interruption of the work or other cause, some of the hot asphaltic cement should be applied to the rebate with a brush, in order to insure a close and powerful cohesion in the joint. In two or three hours after the work is done the road is sufficiently hard to be thrown open to traffic. It is usual to first spread over it a light layer of loamy sand.

The consolidation is sometimes effected with heavy iron rollers of different weights, beginning with the lightest and finishing off with the heaviest, but a roller containing only a single cylinder, if sufficiently heavy to be effective, is apt to cause the soft asphalt to rise up in front of it, and even to tear it asunder as the temperature becomes lower. Two or three rollers, placed close together in the same frame, one behind the other, would be less objectionable.

The method of consolidation with rammers is believed to be the best, all things considered.

It is important that the surface should be evenly finished to the required grade, and be free from elevations and depressions. If otherwise the wear will be more severe, in consequence of the shocks and blows created whenever a wheel rises and falls upon the uneven surface. Moreover, rainwater and all noxious fluids should drain off freely, which they could not do from a rough pavement, and as the asphalt is impervious to water, they would stand in pools until cleaned off or removed by evaporations.

There can, of course, be no vertical absorption from rains, through such a covering. There will also be no lateral absorption if suitable drainage has been provided. The road bed will therefore remain dry at all seasons of the year, and no upheavals from frost need ever be apprehended.

## Asphalt Pavements without Bituminous Limestone.

When it is desired to construct an asphalt pavement, without using the bituminous limestone from the Jurassic region which, as already described, contains both the matrix or binding material, and the body cemented together, and which after disintegration by heat, or by grinding, is capable of being re-united under a new form, and especially in the form of a monolithic sheet suitable for street coverings, there are three essential points demanding consideration, viz :

*First.* To obtain a suitable bituminous or asphaltic cement.

*Second.* To obtain a solid material in the form of

powder, fit to replace the amorphous carbonate of lime of the natural asphalt rock.

*Third.* To combine these two in such manner that they will answer the purpose of a pavement. These points will be briefly considered in their regular order.

## The Asphaltic Cement.

*First.* The imported mineral tar, after proper treatment to adapt it to our climate, by adding a small quantity of refined asphaltum, is a good asphaltic cement. When, however, it is desired to manufacture a suitable cement from the crude materials, the variety of bitumen known as asphaltum or asphalt is employed as the basis.

The asphalts from different localities differ very much in the proportion of asphaltine and petroline which they contain. Too much asphaltine renders the cement brittle in cold weather, while it will become too soft in summer if it contains an undue quantity of petroline. No natural asphalt, whether liquid or solid, has yet been found suitable for all climates and seasons, and it is necessary to mix two or more together, in order to arrive at satisfactory results, having first ascertained the standard quality of each.

The asphalts in common use in this country are derived from deposits found in the islands of Cuba and Trinidad; from Ritchie county, West Virginia, and from the province of New Brunswick. The two last named, known respectively as Grahamite and Albertite, are pure asphalts, containing little if any foreign impurities, while the others contain from 20 to 30 per cent of deleterious refuse matter, which has to be separated by a careful process of refining. They all differ in the amount of asphaltine which they contain, there being the most in Grahamite and Albertite, less in the Cuban, and least in that from Trinidad. There is a liquid asphalt or mineral tar, brought from the isthmus of Tehuantepec, which contains less asphalt than the Trinidad. It contains about 2 per cent of sulphur and a large proportion of water, as received in the market, from which it has to be separated by refining.

These natural asphalts, having each been brought by refining to a standard point of specific gravity and purity, may be used separately, or two or more of them may be mixed together, as the basis of the asphaltic cement. It will be necessary in using only the solid asphaltums to add, at a suitable temperature—say from 250° to 300° Fahrenheit—a small per centage of the proper solvent, such as the heavy oils or "still bottoms," produced in the process of refining crude petroleum, in order to render the cement fit for pavements, and enable it to withstand the changes of temperature in the locality where it is to be used. The precise quantity of heavy oil that it would be necessary to add to any one asphalt, would not be likely to suit another, or the mixture of two or more of them. Chemical research and experiment can alone determine this point. A liquid bitumen containing asphaltine and petroline in the requisite proportions to produce the desired result may replace the still bottoms.

A serious impediment to the successful use of asphalt for pavements in this country, is the condition in which they are shipped from the island of Trinidad and other localities where the deposits are found.

The operation of refining—improperly so called—at the mines, is very imperfectly done, so that variable and sometimes large percentages of refuse and deleterious matter, such as earth, vitrified sand, cinders, ashes, etc., are contained in the material as supplied for use. Dr. Ure, in several analyses, states that he found from 20 to 30 per cent of these impurities. This variable quality of the asphalt, not only seriously impairs its value, but likewise renders the results of its use uncertain, when fixed rules as to proportions, which have been established for an article of standard value, are adhered to.

The manner in which the crude material is treated for the purpose of separating these foreign substances, and distilling volatile matter, is faulty. The heat is often applied in an irregular and unsystematic manner, and with such carelessness as frequently to destroy by overheating much of the most valuable part of the asphalt, leaving the material in such condition, that when employed for pavements, even after the best formula honestly and closely followed, no good results can be obtained.

The preparation of asphaltic cement, suitable for carriage way pavements, requires improved machinery, scientific supervision, and systematic labor, all of which will necessarily enhance the first cost of the article. None of them can, however, be safely omitted.

Any process which will not, with reasonable certainty, secure the uniform application of heat to, and proper supervision over, the distillation and refining necessary to bring the material to its proper condition, will be likely to end in more or less complete failure, for one or the other of the following reasons, namely, (1) it will either result in the destruction of some of the valuable properties of the bitumen, or, (2) it will impair its fitness for pavements by the imperfect evolution and distillation of volatile or other injurious or useless matter, such as water and light volatile oils.

The asphaltic cement when properly prepared must be soft even in cold weather, must give by distillation a hydro-carbon oil of a fire test not less than 200° Fah., must contain no water or any ingredient soluble in water or the urine of animals, and must contain nothing that is oxidizable, or that can be affected by the elements to any serious degree. Changes of temperature will not render it very brittle in winter and soft in summer, even in the variable climate of the United States, although the formula that would suit Boston or New York, would not answer in either New Orleans or Key West, more asphaltine being necessary in the cement intended for the warmer climates.

In striking contrast to the material above described, in every conceivable respect, are all those mixtures of wood-tar, coal-tar, pitch, rosin, etc., which have brought such unjust disrepute upon asphalt pavements.

## The Powder.

*Second.* The specious and altogether worthless street coverings, erroneously called asphalt pavements, to which brief reference has already been made, have owed their uniform failure, not so much to the sand, gravel, lime, etc., employed to give them strength and body, as to the entire absence of a suitable cementing medium.

A fair pavement can be made by properly mixing any kind of fine sand with the asphaltic cement above described, but nothing—not even the pure amorphous carbonate of lime contained in the natural asphalt rock—would make a pavement of any value, if coal or wood-tar, pitch or rosin, or any combination of them, furnishes the matrix.

The best material yet found in the natural state, for the body of the mixture, is very fine sand, carefully screened from all coarse particles and gravel, and composed partly of calcareous matter in which the calcareous ingredient exists mostly in the form of powder. This sand, when rubbed between the fingers, does not convey the idea of being sharp and gritty, but feels something like fine Indian meal. It should contain 25 to 30 per cent of calcareous matter, and preferably 40 or even 50 per cent.

When such a material cannot be found, it can be produced by mixing fine silicious sand and pulverized marl, and where marl is abundant it can be used alone, being first ground to a fine powder. The sand should be as absorbent in character as can be found, and the ratio of absorption among different kinds will, as a general rule, vary directly with the quantity of porous calcareous particles which it contains.

## Mixing the Cement and Powder.

*Third.* The sand and the asphaltic cement are mixed together at a temperature of 250° to 300° Fahrenheit—not higher—in the proportion by weight of 80 parts of sand to 20 parts of cement. The two ingredients are heated separately.

In conducting operations on a small scale, the cement may be heated in an ordinary furnace kettle, and the sand in a broad shallow sheet-iron vessel, so arranged that a fire can be maintained under it. The hot sand is added to the cement in small quantities, accompanied by frequent stirring, and when a homogeneous mixture has been obtained in the required proportions, it is ready to be applied upon the street, in the manner already described for pavements of compressed asphalt rock. Calcareous sand suffers injury if heated much above 300° Fahrenheit, so that a watchful care is necessary in this part of the process.

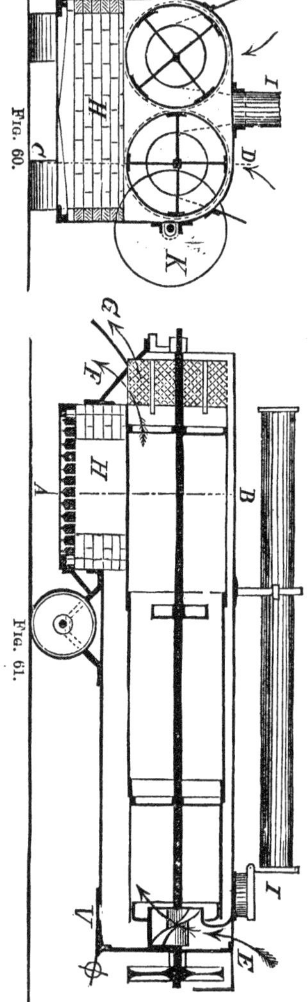

Fig. 60.

Fig. 61.

For extensive work better appliances for heating and mixing the materials than those above indicated would be economical. For the sand, a hollow iron cylinder about 2 feet in diameter and 10 to 12 feet long, set horizontally in a furnace and revolving slowly, or two such cylinders placed side by side, has been found to answer very well.

Such a sand heater is shown in Figs. 60 and 61. Fig. 60 is a section on AB of Fig. 61, and Fig. 61 is a side view and section on CD of Fig. 60. E is the inlet for sand; F outlet for screened sand; G outlet for screenings; H fireplace; I the chimney temporarily lowered; and K the driving pulley. The sand is conveyed from the inlet E into the revolving heater by an archimedian screw. At V is a valve, lightly counterpoised so as to open and allow any sand to escape which may get into the furnace flue S, which surrounds the heater. The whole machine is mounted on a pair of wheels so as to be portable.

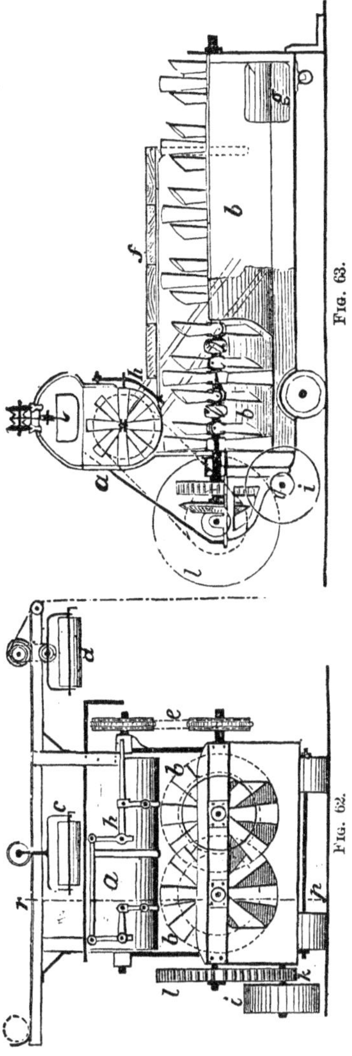
Fig. 63.

Fig. 62.

The asphalt mixer in which the melted cement is incorporated with the hot sand, consists of a horizontal twin pug-mill 6 to 8 feet in length. The two horizontal shafts are parallel to each other, and about 1 foot 9 inches apart, each provided with arms 14 or 15 inches long, set at right angles to the shaft. The arms from the two shafts therefore overlap each other about 7 inches. The bottom of the pugmill is composed of parts of two cylinders of boiler iron, respectively tangent to the ends of the two sets of shaft arms.

Directly over this pugmill is a top mixer, in which the melted cement and hot sand are first mixed together, and into which they are measured in the proper proportions, with vessels suspended above on movable pulleys. See Figs. 62 and 63. Fig. 62 is an end view; Fig. 63 a side view and section on *p r* of Fig. 62; *a* top mixer; *b* twin mixer; *c* asphalt measure; *d* sand measure and hoisting apparatus; *e* driving chain for top mixer; *f* wooden platform for workmen; *g* fire room; *h* fixtures for operating sliding doors in bottom of top mixer for supplying the twin mixer; *i* driving wheel; *k* pinion, and *l* driving cogwheel. When in operation the shafts of the twin mill revolve slowly in opposite directions, the ends of both sets of arms ascending between the two axes, so that their action is to lift the material constantly and let it fall, thus securing a thorough and homogeneous mixture. The whole apparatus is arranged with a furnace underneath.

It is convenient to have both the sand heater and mixer on wheels, so that they can be readily moved from place to place as circumstances require.

## Applying the Mixture.

The asphalt, having been prepared in the manner thus indicated, is in the condition while hot of partly coherent powder, in all respects resembling the powdered bituminous limestone, and in its use, the directions given for laying pavements with that material should be closely followed.

When the foundation is of broken stone, newly laid for the purpose, great care should be taken to guard against subsequent settlement. The stone, or a mixture of stone and gravel, should be of all sizes from 2½ inches in longest dimensions down to fragments not larger than peas, in order to reduce the voids to a minimum. By mixing it with one-fourth to one-third of its volume of some of the best coal tar mixtures used for patent sidewalks, heating the mixture, and then compacting it by rolling or ramming in two layers of 3 to 4 inches each, a good foundation, although inferior to one of hydraulic concrete, will be obtained. When necessary to save expense, the use of the coal tar compound may be restricted to the top layer, or even to a surface coating of the top layer, which should be forced well into the interstices during the process of ramming or rolling. But all these devices for cheapening a foundation are certain to impair its strength and solidity.

One of the patent asphalt roadway pavements is described to consist (1) of a foundation, 5 to 6 inches thick, of broken stone heated and mixed with a bituminous compound, topped off (2) with a thin binding layer of small gravel and stone, over which is placed (3) a thin coating of liquid asphalt. Then follows (4) a layer ⅓ to ½ inch thick of asphaltic mastic; then (5) another coating of liquid asphalt; and finally (6) a top layer 1¼ to 1½ inches thick of asphaltic mastic. It is needless to say that if the nomenclature here employed be correct, and coal tar or some patented compounds are not referred to, this will be a good pavement for footpaths, although unnecessarily thick, but will not be hard enough for carriage ways. By replacing the four top layers—the third, fourth, fifth, and sixth—by genuine asphalt as above recommended for carriage ways, a good street pavement would be obtained.

## Asphalt Block Pavement.

Mention was made on page 177, of the superior toughness, hardness, and incompressibility, conferred on bituminous limestone by compressing it while hot. This property characterizes any genuine asphalt mixture suitable for paving purposes, and advantage has been taken of it, in first forming the material into rectangular blocks under a heavy pressure, and then laying them in courses across the street, substantially after the manner followed in constructing the best stone block pavement. It is, perhaps, needless to say that a pavement of this kind, composed of good materials, properly prepared, and laid upon a firm and unyielding foundation, should be a good one. Specimens of it have been on trial for some years in San Francisco, Cal., where it is styled by the patentee the Imperishable Stone-Block Pavement. The blocks are made with Trinidad asphaltum, softened with 7 to 9 per cent of the heavy oils or still bottoms, used in preparing the asphaltic cement described on pages 182 to 185. This preparation is mixed with hot powdered limestone, or powdered furnace slag, and then compressed with a force of about 50 tons into blocks measuring 4 inches by 5 inches by 12 inches. The pressure, which is applied to the narrowest face of the block, exceeds one ton to the square inch. The limestone or slag is not required to be of the firmness of impalpable powder, but is composed of grains of all sizes from dust up to the size of a small pea.

The blocks are laid close together on their longest edges, in courses across the street, breaking joints lengthwise of the street, the joints being filled with suitable asphaltic cement so as to render the pavement water tight. The foundation should be firm and stable, such as the best of those described on pages 143 to 149. This pavement while new would be nearly as smooth as that of the continuous sheet of asphalt heretofore described, but the wear of heavy traffic would, in a short time, crumble off the edges of the blocks and open the joints at the surface sufficiently to give the horses a foothold, without impairing the imperviousness of the covering. It is suggested that it would be better to form the blocks with slightly truncated or rounded edges, so as to give the requisite foothold when the pavement is laid, rather than to secure the same end by the irregular and ragged abrasion caused by use. As they are homogeneous in toughness and hardness, the blocks can be taken up, when their surfaces become uneven from unequal wear, and relaid in mortar, bottom sides up, with all the smoothness of a new pavement. It may be added that the process of refining and careful manipulation described on pages 182 to 185, is equally necessary whether the material be applied as a monolithic sheet, or as blocks, and any mixture that is suitable for the former is also suitable for the latter ; also, that a form of sand is not a proper foundation in either case.

## Merits of Asphalt Pavement.

The advantages possessed by monolithic asphalt pavements constructed by either of the two methods above described are (1) that they produce no dust and therefore no mud ; (2) are comparatively noiseless, the clicking of the horses' feet excepted ; (3) do not absorb and retain noxious liquids but facilitate their prompt discharge into the side gutters and sewers ; (4) they are impermeable and emit no noxious vapors themselves, or allow their emission from the subsoil; (5) they reduce the force of traction and consequently the expense of wear and tear upon animals and vehicles to a minimum; and (6) although they do not furnish as secure a foothold for animals drawing heavy loads as stone blocks in narrow courses, or as cobble stones, still they do not become polished and slippery from continued wear.

They are adapted to all streets with a grade not steeper than 1 in 48 to 1 in 50, except perhaps those thickly crowded with heavy loads, and liable to be kept constantly wet and slippery during the busiest hours of the day, from the accumulated urine of animals, and where the vehicles are subject to the inconvenience of frequent and sudden halts, starts, and sharp turns, like many of the streets in the lower portions of the city of New York, and especially like Broadway below Fourteenth street, over every transverse section of which, from 12,000 to 14,000 vehicles of all kinds pass daily. In such localities it seems necessary that the pavement should possess that roughness of surface conferred by blocks laid with open joints directly or diagonally across the street.

Where the traffic is of a lighter character, or where there is ample room for conducting it without inconvenience, or where a large proportion of it is pleasure driving, and particularly where the streets are lined with residences on either side, the many advantages of a good asphalt pavement, its cleanliness, its noiselessness, and its imperviousness to noxious fluids—important features in which it stands unrivaled—should not be lost sight of.

Another consideration demanding the exercise of sound judgment is that no pavement combines the opposite requirements of an even surface for the wheels, and a suitably rough one for the horses to travel upon, and a compromise of advantages must therefore be made in most cases.

The wear of asphalt is quite small, its diminished thickness under traffic being principally due to compression and not to abrasion.

It has been ascertained in London that some specimens of Val de Travers asphalt, subject to four years' wear in a street of the greatest traffic, had diminished about one-ninth in thickness, while its specific gravity had increased in about the same ratio.

Some of the same material after fifteen years' wear, in Rue de Bergère, Paris, was found to have lost 12½ per cent of its thickness, but only 5 per cent of its total weight.

This material is, to a great degree, a non-conductor of vibration and of sound. It is much less sonorous than granite, and gives out very little noise from wheels. When properly laid, loaded vehicles make no impression upon it. A four-wheeled truck weighing three tons, carrying a boiler weighing twenty-one tons, passed over a piece of asphalt pavement in Fifth Avenue near Twenty-fourth street, New York, without leaving any mark. This occurred in the last week in June.

The asphalt covering rarely exceeds two or three inches in thickness, and readily adapts itself to any subsidence or movement of the surface on which it rests. A firm and unyielding foundation is therefore an indispensable requisite of a good asphalt pavement. Concrete is recommended for this purpose, in all localities.

As the asphalt surface is smooth and even when properly laid, offering but a trifling obstacle to surface drainage, a comparatively flat cross section is admissible, and, as a precaution against horses falling, may be regarded as indispensable. The inclination toward the side gutters should not exceed 1 in 60. For the same reason the grades must be kept low, preferably not steeper than 1 in 50, in order that great extra effort may be avoided either in pulling on the up grades, or in holding back on those that are descending.

Streets with an undulating grade should not be paved with asphalt upon those portions steeper than 1 in 50.

With suitable appliances, labor at $2.00 per day, natural American hydraulic cement at $1.50, or Portland cement at $4.00 per barrel, and refined asphaltic cement at 1½ cents per pound, an asphalt pavement of the kind described on page 183, and following, can be laid for $2.70 to 2.80 per square yard, exclusive of profit to the contractor, as follows:

| | | |
|---|---|---|
| Concrete foundation 6 inches thick (labor and material), | $1.40 | to $1.45 |
| Asphalt covering 2 inches thick do do..... | 1.30 | to 1.35 |
| Total........................................ | $2.70 | $2.80 |

This includes nothing for earth work, but supposes the road bed to have been suitably excavated or filled in, as the case might be, in readiness for the minor adjustments of grade and cross section, preparatory to laying the foundation.

If the natural asphalt rock from Seyssel or Val de Travers be used for the covering, 80 to 90 cents per square yard will have to be added to the foregoing prices. In the city of Washington, these pavements of the natural rock have cost, within the last six years, $4.25 per square yard inclusive of profit, although the payment was made in depreciated securities.

Among intelligent people, there are many, more or less familiar with bitumen and its uses, who claim that no asphalt pavement can be produced, equal to that made with natural asphalt rock, because that rock *is* natural. This assumes that nature leaves her work so nearly perfect, that improvement upon it is impossible, a premise that cannot be maintained. The crude petroleums possess very little value until they have passed through the laboratory of the practical chemist; and the mineral tars of the Jurassic region, when used for enriching the meagre bituminous limestone, or when added to that material in the manufacture of bituminous mastic, are always refined beforehand. The asphaltic cement to which reference has been made on pages 182 to 185 is prepared from the natural bitumens, and is identical in chemical composition, in color, tenacity, and all other physical features, with the best refined mineral tar, with this advantage over the mineral tar contained in the asphalt rock, that its consistency may be adapted to any climate, while the other is suited to but one. What is wanting in this respect can, however, be easily added for any latitude.

Whether the amorphous pulverulent carbonate of lime, which forms the body of the asphalt rock, can be dispensed with and its place supplied from other sources, is that branch of this subject, to which the differences of opinion really attach. Until quite recently—say within the last eighteen months—the writer somewhat mildly took the negative of this question, and his views were published in a short article on *bitumen*, in Johnson's New Universal Cyclopædia. But the success of certain asphalt pavements in the city of New York, notably those in Fifth avenue near Twenty-fourth street, and in Eighteenth street east of Fourth avenue, leaves no doubt in his mind that the bituminous limestone, and the mineral tars of the Jurassic region can both be dispensed with in the future, for paving purposes. The two pavements referred to were not laid upon proper foundations, but directly upon the old block-stone and cobble-stone pavements set in sand. There has therefore been unequal settlement in the foundations to some extent, to which the sheet of asphalt has conformed, as it always will. In other respects, and indeed in all respects, involving the merits of the material in question, these specimens of street covering compare favorably with the asphalt-rock pavements of Paris and London, with which the writer is to some extent familiar both from personal examination, and from the written reports of experts.

## Comparative merits of Wood, Stone, and Asphalt Pavements.

In this comparison only first-class pavements of their kind will be considered. The brick pavement described on pages 168 and 169 is omitted for want of sufficient data with regard to its durability and cost. The asphalt block pavement, in the absence of direct personal knowledge respecting its several merits, is also omitted, although regarded with great favor.

## Durability.

Assuming the foundation to be firm and solid, so that ruts and depressions cannot form upon the surface except from actual wear, a pavement of stone blocks, of first quality as regards hardness and toughness, will possess the longest life of the three, and one of wood blocks the shortest : asphalt lies between the two and very near to the stone, and will fluctuate from this position with the amount and kind of traffic, and the influences of the climate. As a rule wood must be regarded as the least durable. When it begins to go—at the end of two or three years, under heavy traffic—it wears rapidly into deep and numerous ruts, by the crushing of the blocks to their entire depth. Unless the stone be of excellent quality for pavements, it takes the second place in the order of durability, and asphalt the first.

### First Cost.

The absolute cost of constructing the different pavements will of course vary very considerably with the locality. It is believed, however, that with few exceptions, the following order of cheapness will obtain throughout the United States: viz., first, wooden blocks; second, asphalt, on a solid cobble stone foundation; third, asphalt, on a concrete foundation; fourth, stone blocks on a concrete foundation.

### Cost of Maintenance and Repairs.

Under this head the life or endurance is to be considered, and the total expense must extend over and cover a period of time representing that endurance, under the assumption that at the end of that period, the pavement is in as good a condition as at the beginning when it was new. The repairs for the first two or three years will be comparatively trifling, and in some cities, more especially in England, it is customary for the constructor to maintain the pavements in a good sound condition without charge for one, two, and sometimes three years, and subsequently for a longer period, seldom exceeding fifteen years, for a specified sum per square yard per year.

With regard to wood and asphalt, the recorded observations make it certain that although a pavement of wooden blocks is less costly to construct than one of asphalt, not only is its annual cost for repairs greater, but its mean annual cost during its life, inclusive of the first cost, is also greater than that of asphalt. With regard to stone, there is a vast difference in the endurance of hard and tough basalt or trap, and the average granite and gneiss.

In economy of maintenance per year during the life time of a pavement, including its first cost, the hard basaltic trap rocks should be placed first, asphalt second, and wood third, except in localities where wood is very cheap and suitable stone cannot be procured, or is subject to heavy charge for transportation. Under such circumstances stone would take the third place and wood would rise to the first. Where wood and stone are both expensive, or the latter is not of the best quality with respect to toughness, asphalt would take the first position.

## Facility of Cleansing.

Both mud and dust adhere with more tenacity to wood than to asphalt or stone, more especially after the fibres of the former begin to crush and abrade, and the order of merit in this respect will be first asphalt, second stone, and third wood, whether the cleansing be done by sweeping or by washing. It stands to reason that a smooth, even surface can be cleansed more rapidly than one cut up with numerous joints.

## Convenience.

Mr. William Haywood, C.E., of London, in his report "as to the relative advantages of wood and asphalt for paving purposes," made to the commissioners of sewers of the city of London, March 17th, 1874, says that "asphalt is the smoothest, driest, cleanest, most pleasing to the eye, and most agreeable for general purposes, but wood is the most quiet." It might perhaps be better to say that the noise produced by wood is of a different kind, which may be more disagreeable to some persons and less so to others. Stone is the noisiest of all pavements.

The noise produced by wood is a constant rumble, that by asphalt an incessant clicking of the horses' feet upon the street surface, with scarcely any noise from the carriage wheels, while stone gives out a deafening din and rattle from feet and vehicle combined.

On the supposition that the surface is kept clean by either sweeping or washing, the difference in slipperiness between wood, stone that does not polish under wear, and asphalt, is not great, although enough, perhaps, to place asphalt last; while a horse not only falls more frequently, but recovers himself less often and less easily upon it than upon the others, by reason of the joints in the latter, which give a foothold. When the surface is covered with mud, asphalt is the most slippery of the three, and very little mud makes it slippery. It cannot be said to be slippery when very dry, or, if free from mud, when very wet.

In times of snow there appears to be little if any difference in this respect between wood, asphalt, and stone, but under a sharp dry frost, asphalt and stone are generally quite dry and safe, while wood retains moisture and is very slippery.

In the condition in which they are usually maintained, a slight rain adds to the slipperiness of each of these pavements, with this difference that on asphalt and stone this state begins with the rain or very soon thereafter, while the worst condition of wood ensues later. It however lasts longer than upon the others on account of its absorbent nature. With regard, therefore, to the convenience and comfort of those using the street, as well as those living adjacent thereto, the weight of opinion appears to place asphalt first, wood second, and stone third, for all streets except those habitually crowded with heavy traffic, in which case stone would rise to the first place and asphalt drop to the third.

## Hygienic Considerations.

A practical and general recognition of the fact—so well known in the medical profession, and indeed among all ranks of cultured people—that the pavements of a city exert a direct and powerful influence upon the health of its inhabitants, has never been secured. Most people claim simply that a street surface should be hard and smooth without being slippery, and, as a measure of economy, that it shall be durable and easily cleansed ; but they go no further.

The advantages of noiselessness are recognized by many upon various grounds ; by the large majority as simply conducive to comfort, but by few as conducive to health ; while the kind of material used, provided it satisfies the foregoing conditions, and the character of the surface is satisfactory with regard to continuity and impermeability, is far too generally considered to be a matter of small moment.

The hygienic objections to granite, are first its constant noise and din, and second its open joints which collect and retain the surface liquids, and throw off noxious vapors and filthy dust.

In populous towns there is scarcely a moment of silence, night or day. M. Fonssagrives, Professor of Hygiene at Montpelier, says, "I cannot consider such a perpetual vibration of the nerves as harmless even for those who have been born and bred in the midst of the noise. It is certain that it is a very genuine cause of erethism, and to it must be ascribed the prevalence of nervous temperaments and diseases in the large towns. . . I have known a young girl of seventeen years old, suddenly transported from the provinces to a noisy quarter of Paris, show the most alarming symptoms of nervous disorder, which did not subside until she returned to a quieter and less exciting atmosphere. At the periods of a woman's life when she is most subject to nervous maladies, this danger should be most carefully guarded against. And what shall we say of the nerves of children and invalids? If the former are hard to rear in cities which create hysterics at eight years of age, some blame must certainly be laid upon the air they breathe and the moral conditions in which they have been educated; but some part of the evil must be attributed to the influence exercised by noise on these little beings, in whose organization the cerebral predominance is the most marked feature. As for invalids, quiet is of the first importance, and the noise in the streets is the cruelest stumbling block in the way of recovery."

Dr. A. M'Lane Hamilton, Assistant Sanitary Inspector of the city of New York, in an official report dated October 19, 1874, says, "A quiet and noiseless street pavement would advance the health of the population to a great extent. The sufferer from nervous diseases would find relief from the noise of empty omnibuses and wagons rumbling or rattling on the rough stones, in the event of a removal of this nuisance. In fact there would be many more sanitary benefits resulting from a change than I can here detail."

It is not deemed necessary to enlarge further upon this point. The writings of eminent medical practitioners are full of testimony to the pernicious influence of street noise and din upon the health of the population, particularly upon invalids and persons with sensitive nerves.

The noisome and noxious exhalations emanating from the putrescent matter, such as horse-dung and urine, collected and held in the joints of stone pavements, constitutes another sanitary objection to their use in populous towns. Exceptions to wood may be taken upon the same, and even upon stronger grounds, for the material itself undergoes inevitable, and, sometimes, even early and rapid decay, in the process of which the poisonous gases resulting from vegetable decomposition are thrown off.

The joints of a block pavement, whether of wood or stone, constitute, after enlargement by wear, fully one-third of its area, and under the average care, the surface of filth exposed to evaporation, covers fully three-fourths of the entire street. This foul organic matter, composed largely of the urine and excrement of different animals, is retained in the joints, ruts, and gutters, where it undergoes putrefactive fermentation in warm damp weather, and becomes the fruitful source of noxious effluvium. In dry weather this street soil, of which horse-dung is a large ingredient, floats in the atmosphere and penetrates the dwellings in the form of unwholesome dust, irritating to the eyes and poisonous to the organs of respiration. Its damage to furniture, though serious, is unimportant in this connection. In the side gutters and underlying soil the foul matter exists in a more concentrated form, the supply being constantly renewed from the crown of the street, and in many districts, from the filthy surface drainage of backways and alleys peopled by the poorer classes. Is it too much to say that under such circumstances, the infant population, and especially the children of poor people, in large towns, can only be reared under such predispositions to disease, as will constitutionally render them an easy prey to epidemics in maturer years ?

The foregoing are some of the leading hygienic objections to pavements laid in blocks, whether of stone, wood or other material. There are others peculiar to wood alone arising from its decay, its natural porosity, and the spongy character conferred upon it by wear and crushing.

"Impregnation of the wood with mineral matters, to preserve it from decay, may diminish these evils, but nothing as yet tried prevents the fibres being separated, and the absorption of dung and putrescent matter by the wood being continued. The condition of absorbing mere moisture is of itself bad, but when the surface absorbs and retains putrescent matter, such as horse-dung and urine, it is highly noxious. The blocks of pavement with this material are separated by concussion, and are thus rendered permeable to the surface moisture. Mr. Sharp, who examined some blocks taken up for re-pavement, states that he found them perfectly stained and saturated with wet and urine at the lower portions, while the upper portions were dry. Mr. Elliott, a member of the society, and for many years a deputy of the Common Council of the city of London, has carefully observed the trials of new modes of pavement there, and objects to the wood that it is continuously wet and damp. 'Wood is porous; it is composed of bundles of fibres. It absorbs and retains wet, foul wet especially. The fibres of the wood are placed vertically, the upper ends whereof fray out, are abraded and become like painters' brush stumps, and are almost permanently dirty, or they break like the handle of a chisel which has been struck with an iron hammer or wooden mallet.' This fact is beyond all question. Wood is wet or damp, more or less, except during continued very dry weather. Its structure is admirably adapted to receive and hold, and then give off in evaporation, very foul matters, which taint the atmosphere and so far injure health." (Report of P. Le Neve Foster, Secretary, Society for the Encouragement of Arts, Manufactures, and Commerce: London. 1873.)

Physicians assert that hospital gangrene frequently results from washing the wooden floors of the wards with water, and that on shipboard, new or moist timber, between decks, impairs the health of the sailors. Fatal epidemics at sea have been traced to timber that has become saturated with putrescent matter, or wet with bilge water.

Prof. Fonssagrives, of France, says: "The hygienist cannot, moreover, look favorably upon a street covering consisting of a porous substance capable of absorbing organic matter, and by its own decomposition giving rise to noxious miasma, which, proceeding from so large a surface, cannot be regarded as insignificant. I am convinced that a city with a damp climate, paved entirely with wood, would become a city of marsh fevers."

The dust produced by the abrasion and wear of a wooden pavement is regarded by physicians as extremely irritating to the organs of respiration, and to the eyes, and being light in weight, it floats longer in the atmosphere and is carried to a greater distance, than that from any other suitable material in use for street pavements.

The evidence, from a sanitarian point of view, against the use of wood for paving purposes in populous towns, is very strong, but the evils are not developed to the same extent in all localities. Decomposition begins in two or three years in clayey and retentive soils, while it is very considerably retarded and the wood remains habitually drier and emits less effluvia where the subsoil is sandy and porous.

The most characteristic features and properties of asphalt pavement have been briefly summarized on page 176 and it is not deemed necessary to repeat or enlarge upon them here. Professor Fonssagrives remarks that, "The absence of dust, the abatement of noise, the omission of joints—permitting a complete impermeability and thus preventing the putrid infection of the subsoil—are among the precious benefits realized by asphalt streets."

Upon hygienic grounds, therefore, asphalt conspicuously stands first, stone second, and wood third, in order of merit.

The correct inference from the foregoing discussion is that no one pavement combines all the qualities most desirable in a street surface. It cannot be sufficiently rough, or sufficiently soft, to give the animals a secure foothold, and at the same time possess that smoothness and hardness which is so essential to easy draught. The advantages of open joints and entire freedom from street filth cannot exist together, under any reasonably cheap method of cleansing the surface.

A pavement of impermeable blocks, if laid upon a solid foundation, may be constructed and maintained in a water tight condition, by thoroughly calking the joints with suitable material, leaving the surface sufficiently rough and open to obviate the objection to a continuous monolithic covering, but roughness, combined with the requisite hardness, is incompatible with the freedom from noise attainable with some kinds of acceptable street surface.

In order, therefore, to obtain the best pavement for any given locality a judicious balancing of characteristic merits is generally necessary. The best pavement, so far as we now know, for all the busiest streets of a populous city, where the traffic is dense, heavy and crowded, is one of rectangular stone blocks set on a foundation as good as concrete, or as rubble stone filled in with concrete; and the next best is one of Belgian blocks set in the same manner.

The best pavement for streets of ample width, upon which the daily traffic is not crowded, or for streets largely devoted to light traffic or pleasure-driving, or lined on either side with residences, is continuous asphalt for all grades not steeper than 1 in 48 or 50.

If the blocks of compressed asphalt fulfill their present promise, they may be able to replace those of stone upon streets where the latter are now preferable to a sheet of asphalt on account of the steepness of the grade.

It has been urged, as an objection to a concrete foundation, that it is difficult to take up in order to reach the gas and water pipes. This is true only in the sense that good work is not easily taken to pieces. But such a foundation when torn up or deranged from any cause, can readily be restored to its former condition, and the pavement relaid upon it with all its original smoothness, firmness, and stability, conditions which do not obtain with any kind of pavement laid upon a bed of sand or gravel.

# CHAPTER VI.

## SIDEWALKS AND FOOTPATHS.

SIDEWALKS and other footpaths are usually paved with flagging-stone, bricks, wood in the form of planks or blocks, or some variety of concrete in which either bitumen or hydraulic cement is the binding material. Various kinds of artificial stone have been used for the same purpose. Most of the pavements above named are so well known as to need no mention here.

### Concrete Footpaths.

Concrete footpaths should be laid upon a form of well compacted sand or fine gravel, or a mixture of sand, gravel and loam. The natural soil, if sufficiently porous to provide thorough sub-drainage, will answer.

It is not usual to attempt to guard entirely against the lifting effects of frost, but to provide for it by laying the concrete in squares or rectangles, each containing from twelve to sixteen superficial feet, which will yield to upheaval individually, like flagging stones, without breaking and without producing extensive disturbance in the general surface. When a case arises, however, where it is deemed necessary to prevent any movement whatever, it can be done by underlying the pavement with a bed of broken stone, or a mixture of broken stone and gravel, or with ordinary pit-gravel containing just enough of detritus and loam to bind it together.

In high latitudes this bed should be one foot and upwards in thickness, and should be so thoroughly sub-drained that it will always be free from standing water. It is formed in the usual manner of making broken stone or gravel roads, already described, and finished off on top with a layer of sand or fine gravel about one inch in depth for the concrete to rest upon.

The concrete should not be less than 3½ and need rarely exceed 4 to 4½ inches in thickness. The upper surface to the depth of ½ inch should be composed of hydraulic cement and sand only. Portland cement is best for this top layer. For the rest any natural American cement of standard quality will answer. The following proportions are recommended for this bottom layer :

| | | |
|---|---|---|
| Rosendale or other American cement.................... | 1 | measure. |
| Clean sharp sand.................................... | 2½ | " |
| Stone and gravel.................................... | 5 | " |

It is mixed from time to time as required for use, and is compacted with an iron-shod rammer, in a single layer, to a thickness, less by half an inch than that of the required pavement.

As soon as this is done, and before the cement has had time to set, the surface is roughened by scratching, and the top layer composed of

1 volume of Portland cement, and
2 to 2½ volumes of clean fine sand,

is spread over it to a uniform thickness of about 1½ inches, and then compacted by rather light blows with an iron-shod rammer. By this means its thickness is diminished to half an inch. It is then smoothed off and polished with a mason's trowel, and covered up with hay, grass, sand, or other suitable material to protect it from the rays of the sun and prevent its drying too rapidly.

It should be kept damp and thus protected for at least ten days, and longer if circumstances will permit; and even after it is opened to travel, a layer of damp sand should be kept upon it for two or three weeks to prevent wear while tender.

At the end of one month from the date of laying, the Portland cement mixture forming the top surface will have attained nearly one-half its ultimate strength and hardness, and may then be subjected to use by foot passengers without injury.

The rammers for compacting the concrete should weigh from 15 to 20 pounds, those used on the surface layer from 10 to 12 pounds. They are made by attaching rectangular blocks of hard wood shod with iron, to wood handles about 3 feet long, and are plied in an upright position.

Certain precautions are necessary in mixing and ramming the materials, in order to secure the best results. Especial care should be taken to avoid the use of too much water in the manipulation. The mass of concrete, when ready for use, should appear quite incoherent and not wet and plastic, containing water, however, in such quantities that a thorough ramming, with repeated though not hard blows, will produce a thin film of moisture upon the surface under the rammer, without causing in the mass a gelatinous or quicksand motion.

The concrete may be prepared by hand, or in the concrete mixture Fig. 64. Equal care is essential in mixing and compacting the top layer of Portland cement and sand. The mixing should be so thorough that each grain of sand will be entirely coated with a thin film of plastic cement, with very little excess of cement not thus disposed of.

A characteristic property of this mixture when properly and uniformly prepared, is that it does not assume a jelly-like motion under the rammer. Excess of water must therefore be carefully avoided. The cement must be precisely such that the effect of each blow of the rammer will be distinct, local, and permanent, without disturbing the contiguous material compacted by previous blows. If it be too moist the mass will shake like wet clay ; if it be too dry it will rise up around the rammer like sand. In either case the mass cannot be suitably compacted by ramming, and would therefore be comparatively weak and porous after setting.

The Portland cement and sand may be mixed together by hand on a mortar bed, but that process, to obtain thorough and uniform manipulation, would be tedious and expensive. A better method would be by a cubical box of somewhat smaller dimensions than the concrete mixer referred to above. A kind of trituration, or a grinding and rubbing process of mixing gives the best results. This may be easily and inexpensively secured by putting in the box, with the cement, sand, and water, several smooth rounded pebbles weighing 6 to 8 pounds each. After the batch is emptied out upon the platform, these are taken out for further use.*

* The writer's previous publications, viz., "Limes, Hydraulic Cements and Mortars," and "Béton Aggloméré and other Artificial Stones," give full details on this branch of the subject. They are published by D. Van Nostrand, New York City.

When silicious hydraulic lime, like that of Teil, France, can be procured at moderate cost, it can be used with advantage to replace one-third to one-half of the Portland cement, care being taken to so adjust the proportions that the volume of paste produced by mixing and tempering the cement and lime together, shall exceed by about 25 per cent the volume of voids in the sand, as ascertained by the water test.

In laying concrete footpaths in squares or rectangles, the material is spread and rammed between stout planks set and firmly maintained on edge, with their upper edge coincident with the surface of the path, every alternate square being omitted in the first instance; to be subsequently filled in,—say on the following day—after those first formed have become sufficiently hard to sustain without injury, the ramming of the fresh concrete against them. To prevent adhesion between the squares, the edge against which the new material is placed may be covered with whitewash, or a coat of oil. A strip of felt, muslin, or card board interposed between the squares will answer the same purpose, although this device is covered by the Schillinger patent.

One advantage of this kind of footpath, over that formed in a continuous layer, is that the squares can be taken up to get at water and gas pipes, and then replaced without injury. In some cases it may be advantageous to mould the squares under sheds, and then lay them like common flagging-stones, after they have become sufficiently strong to bear handling and transportation. Three weeks will generally suffice for this purpose. It will be found unadvisable to make them larger than three to three and a half feet square.

The Schillinger pavement for footpaths, which is patented

in respect to the method of preventing adhesion between the squares, is formed substantially after the manner above described, with this important exception and defect, that the top layer, which receives all the wear, instead of being mixed with very little water and compacted by ramming, is applied in a plastic condition as a coat of mortar. It is therefore comparatively deficient in hardness, compressive strength and the power of resisting frost. Its want of compressive strength, in particular, was fully proved by experiments in 1871, recorded in the volume on "Béton Aggloméré and Other Artificial Stones," from which the table page 214 is taken. It shows in a marked manner the superior strength of a mixture that can be compacted by ramming; as well as the superiority of Portland to Rosendale cement.

The surface layer of the concrete pavement above described, resembles in all essential respects the artificial stone to which the name béton aggloméré has been given in France, sometimes known as béton Coignet, from M. Francois Coignet of Paris, who first introduced it. In France, however, the silicious hydraulic lime of Teil replaces the Portland cement to a large extent, some of the strongest samples of the stone having been made with 1 measure of this lime (slaked and in powder) ¾ of a measure of dry Boulogne Portland cement, and 4 measures of sand. The compressive strength of this mixture, when 21 months old, was reported by Mr. P. Michelot, ingénieur-in-chief des Ponts et Chaussée to be 7176 lbs. per square inch for one specimen, and 7405 lbs. for another. The specimens were rectangular blocks 3¼ inches deep, 3 inches long and 2¼ inches wide.

| Proportion of Sand and Cement by Measure (dry). | How Mixed. | Crushing strength of Blocks in gross tons. |
|---|---|---|
| Rosendale cement, no sand...... | Not plastic. | 6.25 |
| do. do. do. ...... | Plastic. | 0.90 |
| Portland cement, no sand....... | Not plastic. | 24.55 |
| do. do. do. ...... | Plastic. | 22.32 |
| Rosendale cement, 1. Sand, 1.2 | Not plastic. | 2.67 |
| do. do. do. | Plastic. | 0.45 |
| Portland cement, 1. Sand, 1.7 | Not plastic. | 24.10 |
| do. do. do. | Plastic. | 8.92 |
| Rosendale cement, 1. Sand, 1.8 | Not plastic. | 1.00 |
| do. do. do. | Plastic. | 0.53 |
| Portland cement, 1. Sand, 2.55 | Not plastic. | 12.50 |
| do. do. do. | Plastic. | 8.47 |
| Rosendale cement, 1. Sand, 2.35 | Not plastic. | 1.34 |
| do. do. do. | Plastic. | Went to pieces in water. |
| Portland cement 1. Sand, 3.4 | Not plastic. | 8.00 |
| do. do. do. | Plastic. | 6.25 |
| Rosendale cement, 1. Sand, 3.5 | Not plastic. | 0.45 |
| do. do. do. | Plastic. | Went to pieces in water. |
| Portland cement, 1. Sand, .5 | Not plastic. | 4.46 |
| do. do. do. | Plastic. | 2.23 |
| Rosendale cement, 1. Sand, 4.7 | Not plastic. | 0.40 |
| do. do. do. | Plastic. | Went to pieces in water. |

The table gives the compressive strength of blocks 3½ inches wide, 5½ inches long, and 3 inches thick, the area under pressure being 19¼ square inches. Some of the blocks were made with little water and compacted by ramming, others with plastic, rather over-stiff mason's mortar, firmly pressed into the moulds with a trowel. The Portland cement was made at Boulogne, France. The blocks were 7 days old, having been kept in water 6 days.

## Asphalt Footpaths.

Asphalt sidewalks may be laid after either of the two methods described for pavements for carriage ways, but the thickness of the foundation, if of concrete, need not generally exceed 3 to 4 inches, and that of the asphalt covering may be restricted to from ⅜ to ½ or at most ¾ of an inch.

In compact clayey soils the foundation should rest upon a lay of sand or gravel, 4 to 5 inches in thickness, to secure sub-drainage, and guard against upheaval by frost. The various patented pavements containing coal tar, resin, pitch, etc., will generally answer as a foundation for the asphalt layer.

Asphalt in the form of Bituminous Mastic is also used for paving sidewalks. This mastic may be prepared by heating together, in a covered iron boiler, mineral tar either natural or manufactured, (see page 173 and 182) and certain calcareous, silicious, or earthy substances previously reduced to powder, and it differs from the mixture used for paving carriage ways only in containing a little more of the mineral tar.

The bituminous mastics of Seyssel or Val de Travers are prepared by mixing the bituminous limestone from those localities, previously pulverized by grinding or by roasting, with the mineral tar derived from the impregnated sandstone. In the Seyssel limestone 7 to 8 per cent of tar must generally be added, while that from Val de Travers will seldom require more than 4 to 5 per cent of tar. The tar required for a given quantity of mastic is first heated in the iron boiler, until the liquid begins to emit a whitish vapor. The powdered stone is then added little by little, care being taken not to add it in quantities large enough to cause a sudden lowering of the temperature. The emission of a yellowish or brownish vapor indicates too high a degree of heat, when the fire must be reduced and the mass stirred rapidly, to prevent injury to the mastic by scorching.

For convenience of handling, the mastic is moulded into blocks measuring about 20 inches by 12 inches by 6 inches. When used it is broken up into small fragments and re-melted, 2 to 3 per cent of mineral tar being then added to compensate for loss at the second heating.

The pulverization of the bituminous limestone preparatory to its incorporation with the mineral tar may be effected by either grinding or roasting.

For grinding it is simply broken up into pieces about the size of a hen's egg and then passed through any ordinary mill. The grinding can best be conducted in cold dry weather, as the stone is then less liable to cake in the mill.

For roasting, the stone is first broken up as for grinding, and then gently heated in a covered iron vessel, accompanied by constant stirring with an iron instrument, which causes the fragments to disintegrate and fall into a partially coherent powder.

Bituminous mastic is suitable for paving sidewalks,

cellars, areas, markets, and for covering walls and arches to exclude water, and prevent leakage.

It is extensively used in fortifications for covering the arches of gun-casemates and powder magazines before the earth covering is put on. When employed for pavements it should be laid upon a concrete foundation of sufficient thickness to support, without settlement or other disturbance, the greatest weight likely to come upon it. This thickness will therefore depend upon the character of the underlying soil, but will rarely exceed 3 inches. The thickness of the mastic covering is usually $\frac{3}{8}$ to $\frac{3}{4}$ of an inch. It is applied by spreading it while hot and plastic, with a wooden trowel or spatula, great care being taken to form a water-tight junction between contiguous strips. Before applying the mastic upon hydraulic concrete the latter should be covered with a very thin slipped coat of common lime mortar; just enough to make it smooth.

As bituminous mastic contains more of the mineral tar or asphaltic cement than the mixture for street pavements heretofore described, it is softer than that mixture, at the same temperature, and is never used for paving carriage ways, or where it will be subjected to the continued tread of heavy animals. It is doubtful whether it is as good even for sidewalks, as asphalt applied in the usual way by ramming.

Where sidewalks have vaults beneath them, it is important that the percolation of water from the top as well as from the side walls next the street should be prevented. When the vaults are covered with arches, a layer of bituminous mastic, and even of some of the best coal tar preparations, properly laid over the arches before the earth filling is put on will prevent leakage from the top.

Another method is to keep the arches so low that a monolithic bed of cement concrete, rather rich in good cement mortar, and not less than 4 to 5 inches in thickness over the crowns of the arches, can be put over the entire width occupied by the vaults and the side wall next the street, the top surface of the concrete being finished with a coat of rich cement mortar, at the proper height and slope to receive the pavement.

Another method still is to omit the arches altogether, and span the entire width of the sidewalk with stone platforms 8 to 10 inches in thickness, of which the outer edges take the place of the curb stones, and the top surfaces that of the side-walk pavement. These platforms fit closely together at the edges, which are calked to render them water tight, and they may rest upon intermediate piers or columns, wherever danger is apprehended of their inability to support the greatest weight which may be placed upon or moved over them.

The vault wall next the street, if properly constructed of rich cement concrete in a monolith 15 to 18 inches thick, will exclude the water perfectly. If of brick laid in bituminous mastic, with all the vertical joints compactly filled with the same material, it will also be water tight if only 12 inches in thickness. But if the bricks are laid in cement or lime mortar, the exterior face of the wall should be coated with bituminous mastic, throughout its entire height, special care being taken to secure a perfect junction between this surface and the roof surface. The filling directly against this wall should be coarse sand or gravel, so that any water from the side gutter will promptly run off.

## Brick Footpaths.

Brick pavements, if laid with carefully selected hard-burnt brick, make very good footpaths for streets devoted mainly to residences, or where there is very little loading or unloading heavy goods at the curb. The bricks should be laid on their edges, with their longest dimensions directly or diagonally across the walk, upon a form of well compacted gravel or coarse sand, or preferably upon a foundation of creosoted boards firmly bedded with a uniform bearing on the sand, to the required inclination. The boards prevent the unequal settlement, almost certain to ensue if they are omitted, in consequence of the narrow bearing surface of the bricks.

## Flagging Stone Footpaths.

Flagging stones laid upon a form of sand or gravel, or directly upon the natural soil when light and porous, form, probably, of all the materials above mentioned, the best sidewalk pavement, and, all things considered, give the most general satisfaction, where they can be procured of good quality and at reasonable cost. The North (Hudson) River blue stone flagging has for many years been in extensive demand for this purpose, in cities and towns of the Atlantic States, north of the Carolinas, located upon water routes. It is strong, hard, and durable, does not polish and become slippery under wear, and resists frost and does not break from upheaval by it, unless unusually broad and thin. The quarries yield slabs of any required thickness and superficial area.

## Broken Stone and Gravel Footpaths.

A very good footpath, suitable for parks, and for the

sidewalks of country roads and suburban streets, can be made with broken stone or gravel, or with a mixture of the two, applied in substantially the same manner and to about half the thickness described in Chapter III, for the construction of road coverings.

After the footpath bed has been excavated to the required width and depth, it should be compacted by a garden roller or by ramming, unless the soil be sufficiently firm without it.

If the soil be wet and clayey, or if it be at all infested with springs, a tile drain of 1½ to 2 inches bore should be laid below the reach of frost, under the centre of footpaths, other than sidewalks, and the trench filling above it should be a sandy or gravelly mixture that will allow the water to percolate freely through it. The subdrainage of sidewalks is presumed to be suitably provided for in connection with that of the roadway.

The lower layer of material to the depth of 4 or 5 inches may be small rubble, or field cobble stone, or the refuse of quarries, and of inferior quality. After it is put in place a roller or rammer may be passed over it, and the interstices partially filled up by breaking off the projecting fragments with a hammer ; or the required evenness may be secured by a second thin layer of smaller stones. A surface of suitable gravel, 2 inches in thickness, applied in two layers, with the necessary raking, sprinkling, and rolling, completes the walk. The surface layer should be of small screened gravel, when practicable.

For park walks the transverse form of the surface should be convex, and the sides of the walk, where no paved side gutters are used, should be sufficiently high to discharge the surface water upon the adjacent turf. In some cases it will suffice to take up the sod on either side of the walk for a width of 2½ to 3 feet, and reset it in the form of a shallow trench, called a sod-gutter, provided with suitable outlets—covered or open drains—at the lowest levels, for carrying off the surface water.

If it be found that the surface water is not promptly conveyed away by these means, or if it injures the sod by wearing it into gullies, the walks must be provided with paved gutters, on one or both sides, as circumstances may require. A neat and durable gutter may be formed of small cobble stones, such as can usually be found in gravel pits.

Where an area embraced by a system of park walks is not susceptible of easy and sightly surface drainage, one or more main covered drains should be constructed, with a sufficient number of branches to collect the water from grated silt-basins located in the depressions of the side gutters. The location and size of these drains, will be governed by the configuration of the ground, the kind of soil, and other circumstances of a special and local character.

The extensive use of hydraulic cement concrete recommended in this volume renders it proper that some general directions for its fabrication should be given for the information of those not familiar with its properties. The following is condensed from the fifth edition of the writer's work on "Limes, Hydraulic Cements, and Mortars."*

## Hand-made Concrete.

Each batch of mortar or concrete should correspond to one cask of the cement. In mixing it by hand labor, four

* D. Van Nostrand, Publisher, 23 Murray Street, New York.

men constitute a gang for measuring out and mixing the ingredients, who proceed to the several steps of the process in the following order :

*First.* The sand is spread upon the platform in a rectangular layer about two inches in thickness.

*Second.* The dry cement is spread equally all over the sand. If lime be used as one of the ingredients, it should first be slaked to a powder by sprinkling, and then mixed with the dry cement, before the latter is spread over the sand.

*Third.* The men place themselves, shovel in hand, two on each side of the rectangle, at the angles, facing inwards. Furrows of the width of a shovel, are then turned outward along the ends of the rectangle, until the whole bed is turned. The two men on one side then find themselves together, and opposite the two on the other side, having, of course, left a vacant space transversely through the middle, of double the width of a shovel. They then move quickly to the ends of the wide furrow and turn successive furrows inward, when the bed occupies the same space that it did previous to the first turning. The turning is executed by successively thrusting the shovel under the material, and turning it over about one angle as a pivot. Each shovel thus moves to the middle of the bed, where it is met by the one opposite, when each man moves back to the side, in dragging the edge of the shovel over the furrow he has just turned.

*Fourth.* A basin is formed by drawing all the material to the outer edge of the bed.

*Fifth.* The water is poured into the basin thus formed.

*Sixth.* The material is thrown back upon the water, absorbing it, when the bed occupies the same space that it did in the beginning.

*Seventh.* The bed is turned twice, by the process above described. If required for mason's use, the mortar is heaped up, to be carried when and where required. If for concrete (the mortar occupying the rectangular space as at first).

*Eighth.* The coarse materials (whether broken stone, bricks, gravel, shells, or a mixture of two or more or all of them) are spread equally over the bed.

*Ninth.* A bucket full of water more or less (depending on the quantity of stone, their absorbing power, and the temperature of the air) is sprinkled over the bed.

*Tenth.* The bed is turned once as before, and then heaped up for use. The act of heaping up, when done with care, has the effect of a second turning.

The time consumed in making a batch of concrete, composed of one barrel of cement, two and a half to three barrels of sand, and five or six barrels of the coarse materials, is from twenty-five to twenty-eight minutes. An experienced gang of first-rate laborers can do it in a little over twenty minutes. If lime be added, the amount of sand and coarse materials, and the time required for mixing are proportionally increased.

## Mill-made Concrete.

Mill-made concrete possesses sufficient superiority over that manipulated by hand, to justify the expense of providing suitable power and machinery, when operations of considerable magnitude are to be carried on. The more thorough manipulation secured by machinery, enables a smaller proportion of the cementing substance to be used, and effects a saving in the cost of both materials and labor.

## The Cubical Concrete Mixer.

This mixer, Fig. 64, consists of a cubical box made of hard-wood plank or boiler iron, measuring about four feet on each edge in the interior, rigidly mounted on an iron

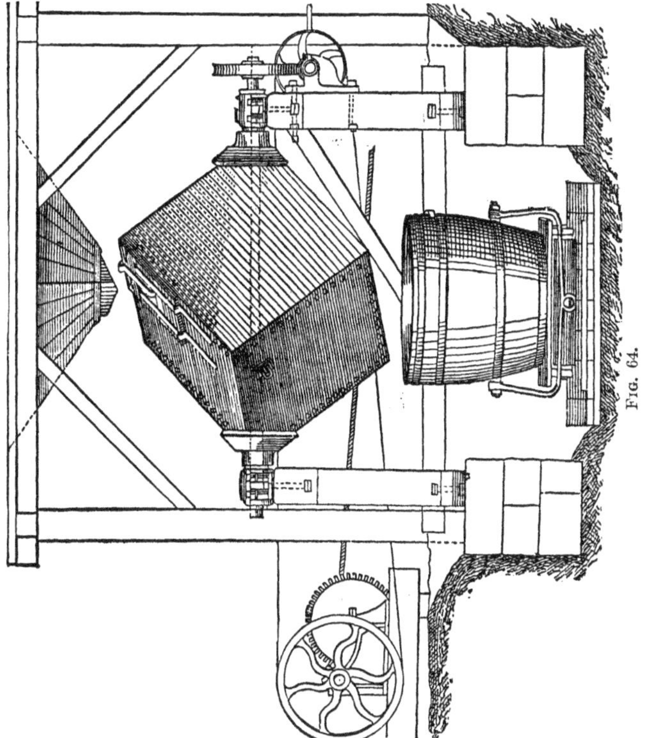

FIG. 64.

axle passing through opposite diagonal corners. It is provided with a trap door about two feet square, close to one of the six angles farthest from the axle, for charging and emptying the box. Eight to ten revolutions of the box, made in less than one minute, are found to be quite sufficient to produce a thorough incorporation of the ingredients. A small steam hoisting engine, which may be used for other purposes at the same time, furnishes the best power for turning the mixer, and screw gearing is probably the best method of applying it.

The mixer is charged through a hopper, by means of a tub, swung from a common derrick crane, and holding just one batch of concrete, the volume of which should not exceed one-third to two-fifths the entire capacity of the box.

The crane should of course be worked by the same power that turns the box, and should have a sweep reaching from the platform where the materials are measured to the hopper.

The process should be conducted in the following order :

*First* and *Second*, spread the sand and the cementing material upon the platform, as in direction for hand-made concrete.

*Third.* The dry materials may be mixed together with shovels, as for hand-made concrete, or they may be only partially incorporated by long teethed rakes passed back and forth through them without disturbing the position of the bed.

*Fourth.* Empty the coarse materials upon the bed of sand and cement and spread them over the same, not necessarily with much care.

*Fifth.* Dash over the bed the requisite quantity of water, in such manner that it will be absorbed by the material, and not run off upon the platform.

*Sixth.* Shovel the materials into the tub, taking care that each shovel full shall contain a portion of each of the ingredients.

*Seventh.* Empty the tub into the box and set the latter in motion.

*Eighth.* After ten or twelve revolutions, occupying about one minute, stop the motion, open the trap-door and empty the mixed concrete into the tub, so that it can be deposited by the crane in some convenient spot within its sweep, and thus be out of the way of the succeeding batch.

It will generally be found convenient to convey the concrete to its allotted place in wheel-barrows. It should be compacted with rammers, in horizontal layers 5 to 6 inches in thickness, until all the coarse materials are driven below or flush with the general surface.

As a rule concrete should be compacted in place before the cement has had time to take its initial set. Where the cement contains quicklime, a delay of a few hours is sometimes necessary to allow the lime to become thoroughly slaked.

## CHAPTER VII.

### TRAMWAYS, AND STREET RAILWAYS.

A horse can draw, upon a good stone tramway, a load 11 times as great as he can move with the same effort and at the same speed upon an ordinary gravel road, the force of draught being only $\frac{1}{179}$ of the load in the first instance while in the second it is $\frac{1}{16}$. Even upon a very dry and smooth broken stone road—*i. e.* a macadamized road in its best condition—the tractive force is $3\frac{1}{2}$ to 4 times as great as upon a good stone tramway.

The marked advantages of a hard smooth surface for the wheels of heavy vehicles to move upon on the one hand, and the comparatively great expense of providing such surfaces on the other, has led to the practice in some localities of restricting the width of the wheel tracks to what will simply suffice for the convenient use of the several kinds of vehicles upon which the traffic is conducted, while the rest of the roadway is finished with a less costly covering.

A construction of this kind is called a tramway, which consists of two parallel tracks of suitably smooth and hard material to receive the wheels, while the spaces between them on which the animals travel, as well as the road surface on either side, is paved with a different material.

The wheel tracks are usually of stone ; occasionally of wood or iron.

As tramways are intended for the equal and common

use of all classes of vehicles, and not, like street railways, for the exclusive benefit of specially constructed cars restricted to one kind of traffic, their construction and maintenance properly belong like street paving to the municipality, rather than to private corporations. They possess certain advantages over street railways in being adapted to every variety of traffic and vehicles, with entire freedom to leave the tram when needful without becoming helpless or inefficient, and return to it as occasion or convenience may suggest. Stone tramways are in general use in Southern Europe, particularly in Turin, Milan, Verona, and many of the smaller cities and towns of Northern Italy.

## The Italian Tramways.

The Italian stone tramways consist of two parallel lines of granite blocks or slabs, each slab being usually about 2 feet in width transversely, 8 inches in thickness, and 4 to 6 feet in length. The blocks are laid end to end with close joints. The clear distance between the two lines is about 2 feet 4 inches, making the width between the two axes or center lines 4 feet 4 inches, which is about the average width between the carriage wheels. The roadway is usually formed with a slight inclination from the sides toward the centre, the tramway blocks being laid to the same inclination, with their upper surfaces flush with the road surface on both sides.

The horse track between the blocks is therefore the lowest part of the road. It is paved with cobble stones from the neighboring streams, forming a shallow concave channel along which the surface water flows away into suitable cross drains.

The wings of the road may be paved, Macadamized, graveled, or left as earth roads.

The foundation for these Italian trams usually consists of a bed of screened gravel 5½ to 6 inches in depth, surmounted with a 2 inch layer of sand in which the granite blocks are set.

The road bed is well compacted by ramming or rolling before the gravel is spread, and this is also watered and rolled or rammed in the usual manner. Sub-drainage is provided in soils which require it.

The surface drainage is discharged by the central gutter between the trams, into sub-drains, through vertical shafts covered with stone gratings. The gratings are formed from a single piece of granité, cut concave on the top to correspond to the surface between the trams, and usually provided with three slots, each about 12 inches long, 1½ inches wide, and 8 to 10 inches apart.

The cost of constructing one mile of the tramway above described, with wages varying from 3 to 3½ francs per day for stone cutters, 1½ to 2 francs for common laborers, and 2 francs for pavers, amounts to about $8,600, gold. This includes the paving between the trams, the subdrains, and the openings in the central gutter leading thereto covered with granite gratings, and surface grooving the blocks to give horses a foothold on the trams when turning out on steep grades. It does not, however, include the cost of paving the roadway outside the trams.

Although the first cost of a good stone tramway is comparatively large, it possesses a long life, and the necessary annual expense upon it for repairs is but a mere trifle. It is an error to assume that they are out of date, although their usefulness and their general adaptation to the necessities and conveniences of traffic, have been greatly restricted by the introduction of steam and street railways. They are certainly not adapted to the most crowded streets of a city, and as a connecting route between neighboring towns, a railroad, although costing three to four times as much, might in most cases be preferable ; but upon wide streets and over short suburban lines with a large traffic, and in special cases in manufacturing and mining districts, and in large cities, as a connecting link between the termini of railroads, where steam cars are inadmissible, they are able to supply a convenient and inexpensive method of carriage, while the current outlay for maintenance, which amounts to a heavy tax upon all other kinds of roads, is only nominal.

Where the haulage is heavy both ways the tramway should have a double track, the centre line of blocks being common to both, and wide enough to allow the vehicles to meet and pass each other without leaving the trams.

To insure a greater degree of permanence and stability, the blocks should be bedded in hydraulic mortar, upon a concrete foundation 6 inches thick. This would allow a reduction in their thickness to about 6 inches.

The horse track should be paved with stone blocks as greatly preferable to cobble stones.

An excellent substitute for a stone tramway is a stone block pavement of carefully selected blocks laid in mortar upon a concrete foundation. Its width for the vehicles in common use need not exceed 13½ to 14 feet to enable them to meet and pass each other without collision. It should be along the middle of the roadway, and slope each way from the crown at the rate of about 1 in 40, the wings being finished with any covering suitable for the neighborhood.

Tramways should of course be laid on the grade of the street as usually constructed, and there can be no advantage in having the middle of the street lower than the sides.

## Street Railways (Called also Tramways).

On a well built railroad, the force required to move a car upon the level rail, at a speed of 5 miles per hour is not far from $\frac{1}{230}$ to $\frac{1}{250}$ of the total weight of car and load, varying within these limits with the state of the rail with respect to moisture and dryness.

The following rule is the one in common use for obtaining this resistance :

1. Multiply the weight in gross tons by 6. The product regarded as pounds will be the friction.

2. Multiply the weight by the velocity in miles per hour and divide by 3. The quotient will be the allowance to be made for concussion, in pounds.

3. Square the velocity in miles per hour, and multiply the square by the frontage in feet, and divide the product by 400 for the resistance of the atmosphere in pounds.

4. The sum of these three results is the total resistance in pounds.

Upon street railway lines, in consequence of the presence at all times of more or less dust and stiff mud upon the rails, the tractive force is comparatively large. In the average condition of the road it may be set down as fully $\frac{1}{120}$ of the loaded car, so that a car weighing 4,000 lbs. carrying 28 passengers, each weighing 150 lbs—total 8,200 lbs—would require the exertion of a force of $68\frac{1}{3}$ lbs ($\frac{8200}{120}$) to move it on a level rail at a low speed. Upon descending grades of 1 in $68\frac{1}{3}$ the brakes would not therefore have to be applied.

In practice the grades must conform to those of the street, and for short lengths may be even steeper than would be suitable for ordinary vehicles upon a good street surface. The question of grades, therefore, for street railways, except in special cases, resolves itself into the adoption of those already existing.

Upon the Bleecker street and Fulton Ferry line, in New York city, there are two very steep grades, one with an aggregate rise of 10.3 feet in 225.25 feet, equal to a mean of 1 in 21.8, and the other with an aggregate rise of 10.3 feet in 248.8 feet, equal to a mean of 1 in 24. The point of highest grade upon each ascent is very considerably steeper than the average rise, which includes those portions with a gentle grade near the foot and the crest of each stretch.

## Drainage.

Upon streets suitably provided with paved carriage ways and sidewalks, and with sewers, there is no occasion to make any special or additional provision for the drainage of street railways, but for lines located upon country or suburban roads, the same precautions must be taken to secure thorough surface and sub-drainage that have already been described as necessary for ordinary roads. There is no occasion to repeat them here.

## Construction.

A street railway, as almost universally constructed in the United States and elsewhere, consists of rolled-iron rails, laid upon longitudinal timbers or stringers, resting upon timber cross ties. The top of the rails should set flush with the surface of the street, and they should, preferably, be of such patterns and laid to such gauge as will least incommode the ordinary traffic conducted with the vehicles of the neighborhood, for it will rarely occur that the interests of the railway, and those of the truck, cart, and express wagon will be other than identical. Upon crowded streets in particular, and generally in the business portions of cities and large towns, every device calculated to keep the current of traffic moving, and prevent blockades, is a benefit alike to all.

## Rails.

It is desirable that the car wheels should bear upon the rail directly over the centre line of the stringers, or as nearly so as possible, in order to obviate any tendency of the rail to cant to one side, when the wood begins to become soft and weak from decay. This condition however would exclude the rail with the single rib, raised on one side only (See Figs. 68 and 69) and having a broad flat surface occupying the rest of its width, which is really the form offering the least interference with the traffic conducted on ordinary carriages ; for the broader the surface upon which such carriages can track, the less will be the difficulty, and the less the wrench upon the wheels, as well as upon the rails, in taking and leaving it.

In some cities the pattern of the rail, as well as gauge of the track, is prescribed by municipal authority, with special reference to obtaining such a railway as will not only reduce to a minimum the annoyance occasioned by the rails to promiscuous traffic, but will enable them even to contribute to its promotion and convenience.

A grooved rail, as a general rule, is not the most desirable either for the railway company, or the ordinary vehicles upon the street. It collects the dust and mud, and, in cold weather, gets filled with ice and snow, thus greatly augmenting the tractive force of the car; while the wheels of wagons and hacks, and especially of all of the lighter and frailer classes of carriages, having once entered the grooves, experience a severe strain, and are not unfrequently twisted off, in leaving them, while the rails themselves are thereby more or less disturbed and in time loosened at their fastenings. In Fig. 65, serviceable and convenient forms are shown at **b**,

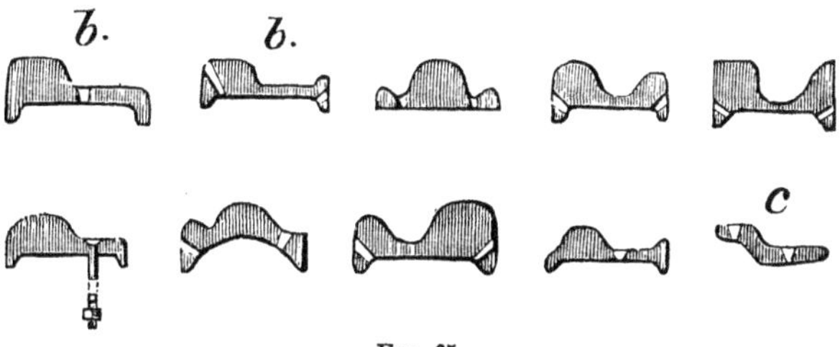


Fig. 65.

which if made of 55 pounds weight per yard will answer for heavy traffic. For light traffic form **c** of 30 to 35 pounds weight has been found to be suitable. Rails are sometimes made of 70 and even 75 pounds weight to the yard. Cast iron rails have been tried without giving satisfaction.

## Stringers, or Sleepers.

The purpose of the stringers is two-fold: to secure a uniform bearing for the rails, and to raise them to the level of the street surface. They may be of southern or of ordinary white pine, preferably the former, on account of its superior stiffness and hardness. Any other kind of durable wood possessing these qualities, may of course be used, provided its price be such as not to exclude it. The sleepers are sawed, the usual dimensions being 7 to 8 inches

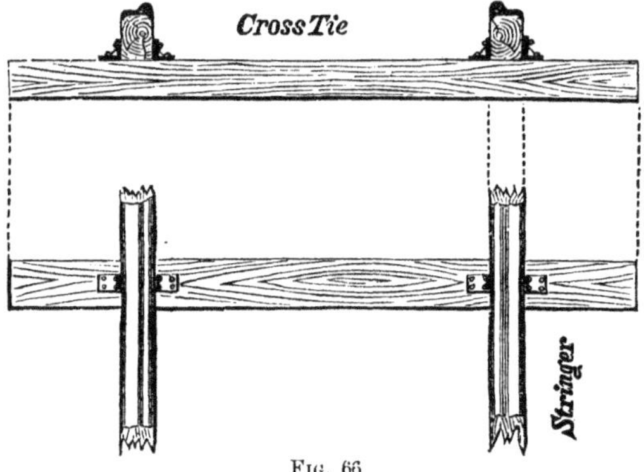

FIG. 66.

in depth, with a width equal to that of the rail, and a length varying from 25 to 40 feet. (See Figs. 66 and 67.)

## Cross Ties.

The cross ties (Fig. 66) may be of any durable wood, either white or yellow pine, chestnut, or white oak, hewn or sawed, 6 to 7 inches wide, 5 to 6 inches deep, and of such

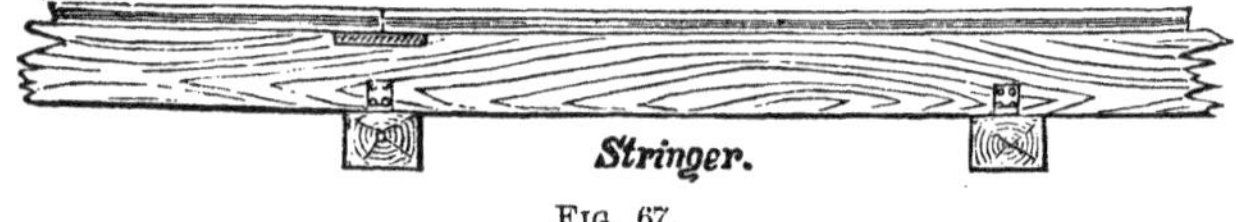

FIG. 67.

length as to reach about 12 inches beyond the stringers on either side. They may be faced on the top and bottom only, the bark alone being taken from the sides.

Upon streets suitably sub-drained, the cross ties are simply laid in trenches excavated to receive them, care being taken not to loosen up the soil to a greater depth than is requisite to give them a firm and level bearing throughout their entire length, with their top surfaces parallel with the line of the grade. The earth should be packed around them and under their edges by tamping with a crow-bar or other suitable implement, to guard against subsequent settlement or disturbance.

When suitable sub-drainage has not been provided, as will often be the case upon suburban streets and country roads, and especially where the railway is to rest upon clayey or spongy soils, the bed should be sub-drained, in substantially the same manner prescribed for ordinary roads, by excavating to a width exceeding the length of the cross ties by at least some inches, and to a depth of 6 inches below them. The trench thus formed, after suitable cross drains have been constructed, is then to be filled in with a ballast of broken stone, gravel, coarse sand, or a mixture of them all, which should be thoroughly compacted by ramming in layers, to guard against further settlement or shrinkage. When the filling has reached the requisite height to receive the cross ties these are placed in position, the material under their edges further compacted by tamping, and the filling continued to the level of their top surfaces.

The cross blind-drains should be at least 12 to 18 inches in depth, below the bottom of the filling or ballast, and should extend out on either side to the side ditches.

To preserve them from early decay, the stringers and cross ties should be creosoted, in the manner described on page 162 which, if thoroughly done, will add at least ten years to their life.

## Fastenings.

No little difficulty has been experienced in firmly securing the rails to the stringers, and the stringers to the cross

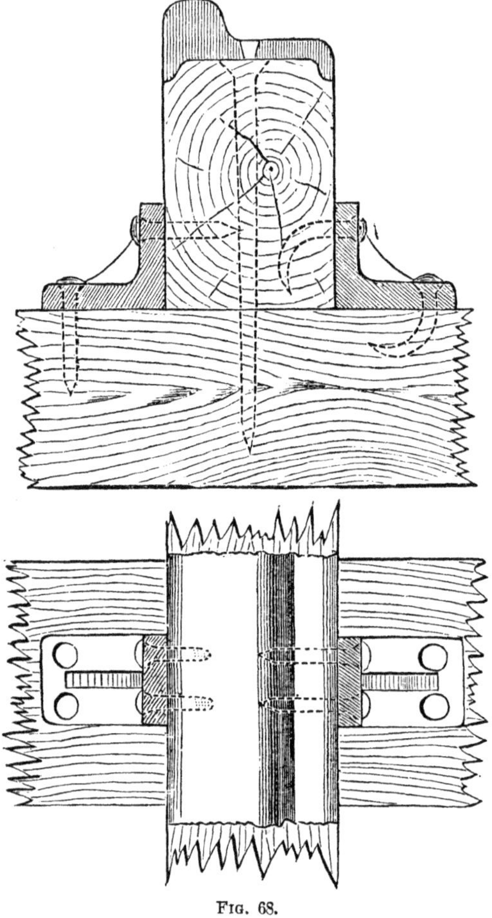

Fig. 68.

ties, and no system or method of fastenings has yet been

devised, possessing so few objectionable features as to command general adoption. It has been found impossible to do away with the use of spikes, bolts, or pins, and these become loose from the gradual enlargement of the holes occasioned by incessant vibration.

For securing the stringers to the cross ties, square or octagonal oak or locust pins have been used, and they possess the advantage that the spikes for fastening the rails can be driven into the pins, in cases where they happen to come directly over them. Half inch round iron bolts, or long half inch square spikes, are generally preferred to wooden pins. The ends of the stringer with an iron plate under them, should rest on a cross tie, and not fall between them, but those on opposite sides should not meet on the same tie.

To prevent the spreading of the track, cast iron knees or angle irons are spiked to the stringer and to each tie or each alternate tie. In some cases these knees are placed on both sides, in others on the outside only of each stringer. (see Fig. 68). Transverse wrought iron tie bolts, of about ½ inch round iron, passing through both stringers, and having a head on one end and a screw and nut on the other, have also been used for the same purpose either with or without the knees. Such ties should be placed near the bottom of the stringers, in order that they may receive no injurious tension or strain from the sinking of the pavement above them.

Additional security against spreading may be given by a good block stone pavement, upon the entire street, including the space between the rails, provided the stones be set firmly against the stringers on either side. Even where a cheaper kind of pavement is used for both horse track and street, it is desirable that one row at least of stone blocks should be set on each side of each stringer, and these should be composed of alternately long and short blocks (as shown in Fig. 69), so as to tooth into the contiguous pavement, and thus avoid a continuous joint which would wear into a rut.

The best pavement between the rails, and upon which the animals appear to travel with greater confidence and less fatigue than upon any other possessing the requisite firmness

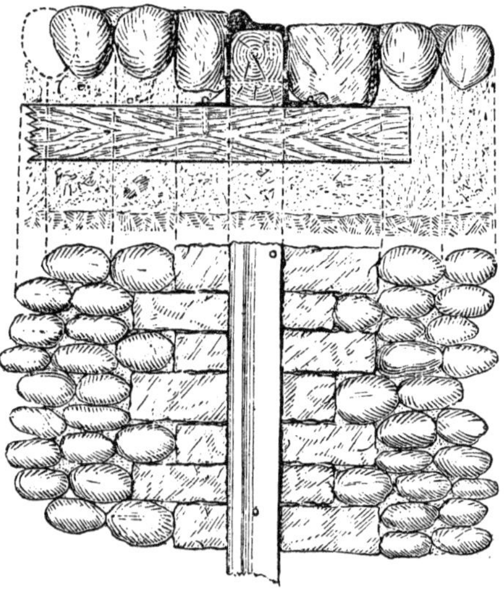

FIG. 69.

and durability, is one of rather small cobble stones, laid with a very slight inclination from the centre toward the rails.

The top of the pavement should be at the same height as the top of the adjacent edge of the rail. With a rail therefore, having a single rib on one side and a single horizontal flange on the other, the pavement next the rib will be as much higher than that next the flange as the rib projects above the flange. If the rib be on the inside of the rail the horse track will be higher than the rest of the street, and if on the outside, it will be lower.

Many methods of fastening the rails to the stringers have been practiced, among which that of clinch spikes with countersunk heads, driven vertically through the bed or thin portion of the rail, well into the sleeper, is about the best. When the width of the bed will admit of it, the spikes should be placed two to three feet apart, alternately near its inner and its outer edges, so as to give the rail a firmer seat, and counteract its tendency to cant to one side or the other. Sometimes the spikes pass through the edge of the rail diagonally, and sometimes screws are used diagonally or vertically. When rails with vertical webs at the sides, reaching some distance down the faces of the stringer, are used, the fastenings may be spikes, screws, or staples passing through the rail into the timber horizontally. The ends of the rails should meet as far as possible from the ends of the stringers.

The perishable nature of wood is a source of serious expense in the maintenance of street railways, and efforts have been made to substitute iron and stone for the timber cross ties and sleepers, but no trial of this method of construction has extended over a sufficient period of time to fully test its practical value.

Many experimental attempts have been made to replace horses by mechanical motive power, for the propulsion of street cars; and gas, hot air, electricity, as well as steam, have been suggested and some of them tried as motors, but never with entirely satisfactory results. In the present condition of the problem, the great desideratum appears to be a small, noiseless, spark-and-smoke-consuming steam engine.

## Car Starters.

The frequent haltings and startings which are necessary for the accommodation of passengers using street cars, operate as a most serious tax upon the endurance of the horses, and within the last few years many mechanical devices have been tried, mostly as matter of experiment, with the object of removing or lessening this evil.

The prevailing idea appears to have been to store up the power necessarily exerted by the brakes in bringing the car to a state of rest, by the compression of some form of spring, and then to expend it in turning the car wheels, so as to assist the horses at the moment they are required to start forward.

It is to be regretted, however, that no car starter has yet been invented, possessing sufficient practical merit to command general approval. They have all failed in a greater or less degree, to render with reasonable certainty, and at the proper moment, the measure of aid required of them. In order to be efficient, they should be under the easy and quick control of the driver, and possess enough initial power to move the car within 1 to 1½ seconds after the signal to start is given, and before the horses have taken the draught upon their collars.

The principle upon which the Crozier car brake and starter operates will be understood from Fig. 70. The upper portion of the figure represents the apparatus in plan, while an end view is given in the middle, and a side view in the bottom cut.

A is a bevel wheel keyed fast to the car axle, and B is another wheel facing it but fitted so as to turn and slide on the axles and having a rose clutch, C, for connecting to the axle so as to turn it when desired; it also has a sleeve D, extending nearly to wheel A, and coupled by a crotched arm E, with a rock shaft below (seen at F in the bottom cut) by which the driver, by means of a lever at W, shifts the gears as required for stopping and starting.

The sleeve D, is also the bearing for one end of the shaft G, on which is a wheel I, for gearing with wheel A, also a wheel H, for gearing with wheel B, and also a drum K, for winding up a spring or springs L, and which has its other bearing in the arm J, of the rock shaft F, so that by the oscillation of the rock shaft through the medium of the lever at W, the shifting is effected.

When the car is moving forward, if the driver by a movement of the lever at W, gears the shaft G, with the wheel A, by means of the wheel I, the drum K winds up a chain which compresses the spring L until the car stops, the wheel B being at this time disconnected with the clutch C, and free to move around. At the signal to start, the shaft G is thrown out of gear with the wheel A, and the wheel B at the same movement engages with its clutch at C. The spring then reacts, and in unwinding the chain from the drum K, starts the car through the medium of the wheels H and B. The wheel B, being larger than wheel A, affords a greater leverage to the power of the spring in starting, than that applied in compressing it when stopping.

## The Fireless Locomotive.

Upon the subject of mechanical motive power for street cars, to which reference was made on page 240, it seems proper to mention an invention of Dr. Lamme of New Orleans, consisting of a small locomotive with a boiler of

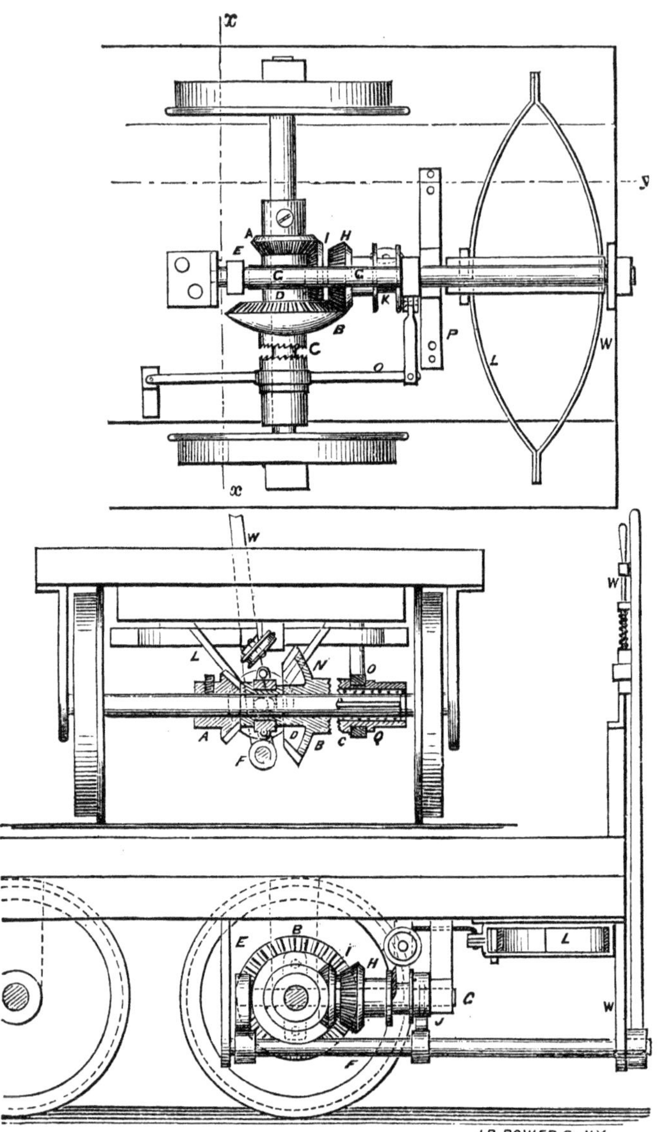


FIG. 70.

about 60 cubic feet capacity, but without any furnace or other appliance for heating it.

The principle upon which it is expected to work may be stated briefly as follows :

"When required to start the car, the boiler is nearly filled with cold water, and the locomotive is run alongside of a large stationary boiler, working at a pressure of 200 lbs. The steam pipe of the stationary boiler is connected with the locomotive boiler, steam then rushes into the latter, and in a few minutes it raises in the cold fireless boiler a pressure of 180 lbs. The connection with the stationary boiler is then uncoupled, and the fireless locomotive is ready for work."

The experiments which were made with this fireless engine were tolerably successful on roads that were level or had only very low grades, and while there were very few stoppages; but upon steep grades and when operating with frequent halts, its limited power was soon rendered inefficient by the constant loss of heat by radiation.

As street railways must adopt the ordinary grades established for other vehicles, and as the stoppages must necessarily be frequent to allow passengers to get in and out, and must be made, even upon ascending grades, unless they are exceptionally steep, Dr. Lamme's locomotive will require extensive modification and improvement before it can be accepted as a practical solution of this question.

There are such strong prejudices in the mind of the general public against placing a steam boiler in a car occupied by passengers, even although it may be kept entirely out of sight, and so arranged as not to interfere with their comfort or convenience, that it is doubtful whether a self-contained steam car can ever commend itself to popular favor.

It is to be hoped, for the benefit of all, that the inventors who do not share this view, will be able by mechanical contrivance and skill, to remove the objections on which it is founded.

## Statistics of Street Railroads.

The following tables, giving the chief particulars of certain horse railway companies, have been arranged at the request of the author, by Mr. Isaac Newton, Engineer, of New York city, from the annual reports of the Massachusetts Railway Commissioners, and the State Engineer and Surveyor of the State of New York for 1873. The examples have been selected so as to give information respecting the operation of such railways in the crowded streets of cities, as well as on ordinary country roads or turnpikes.

The particulars given may be divided in two classes : *first* those regarding the cost of the construction of the roads : *second,* those referring to the cost of maintenance, including all operating expenses, and the amount and source of the revenue. In the case of the New York railways, the figures respecting the cost of construction are not in all cases reliable, but those regarding the operating expenses are, in the opinion of the writer, correct : the latter are obviously of the most importance. Engineers can estimate with all needful accuracy the cost of constructing and equipping a proposed horse railway ; but respecting the cost of operation, the facts obtained from experience with existing roads are the only safe guides to a close estimate.

Regarding the Massachusetts Roads, both classes of figures may be taken with confidence, as accurate statements of the facts.

TABLE I.—*Particulars of Horse Railroads in State of New York,*

| | Second Avenue Railroad. | Third Avenue Railroad. | Sixth Avenue Railroad. | B'way and 7th Av. R. R. | Eighth Avenue Railroad. | Dry Dock and East B'way Railroad. |
|---|---|---|---|---|---|---|
| ***I. Description of Road.*** | | | | | | |
| 1. Length laid........ | 10 miles. | 8 miles. | 4 miles. | 8 miles. | 9,50 m. | 10.73 m. |
| 2. Length of double track, incl. sidings. | 11 miles. | 10 " | 4.375 " | 16.25 m. | .......... | .. ........ |
| 3. Weight of rails, pounds per yard... | 60 lbs.... | 56 to 68lbs | 60 lbs. | 52,62,65 lbs | 50to65 lbs | 48,52,62 lbs |
| ***II. Revenue.*** | | | | | | |
| 1. From passengers.. | $678,547 | $1,512,39 | $737,35 | $894,18 | $757,15 | $776,80 |
| 2. Horses, manure, old material, advertising in cars, and miscellaneous...... | $2,637 | $628,42 | $201,07 | $25,97 | $40.887 | Bonds—$490.900 Mis., $7,915 |
| 3. Total revenue..... | $681,185 | $2,140,82 | $938,43 | $920,15 | $798,040 | $1,275,623 |
| ***III. Operating particulars.*** | | | | | | |
| 1. Number of passengers carried........ | 13,570,955 | 26,950,00 | 14,747,141 | 17,883,776 | 15,143,048 | 15,536,16 |
| 2. Rate of fare....... | 6c. thro. 5c. way. | 6c. thro. 5c. way. | 5 cents. | 5 cents. | 5 cents. | 5 cents. |
| 3. Rate of speed, including stops...... | 1 h. 20 m. | 1 h. 20 m. | 6 m. p. h. | 46 min. | 90 min. | .......... |
| ***IV. Equipment.*** | | | | | | |
| 1. No. horses or mules | 1022 | 1841 | 1,097 | 1,146 | 1,000 | 835 |
| 2. Number of cars... | 154 | 251 Pas. 11 Frt. | 100 | 141 | 110 | 131 |
| ***V. Cost of Road and Equipment.*** | | | | | | |
| 1. Road-bed super-structure, incl. iron, | $1,816,412 | $1,500,000 | $857,444 | $2,841,270 | $844,459 | $340,241 |
| 2. Land, buildings, & fixtures, incl. land damages........... | $410,593 | $1,817,365 | $827,590 | $657,360 | $599,211 | $474,830 |
| 3. Horses & harness. | $173,187 | $250,000 | $385,817 | $191,050 | $102,390 | $141,775 |
| 4. Cars.............. | $111,550 | $190,000 | $111,103 | $157,478 | $137,513 | $134,000 |

*compiled from Report of N. Y. State Engineer and Surveyor for* 1873.

| | 42d and Grand St. Ferry Railroad. | Bleecker St. and Fulton F. Railroad. | Ninth Avenue Railroad. | Brooklyn Bath and Coney Is. Railroad. | Brooklyn City Railroad. | Brooklyn City and Newtown Railroad. | Coney Is. and Brooklyn Railroad. | Buffalo Street Railroad Company |
|---|---|---|---|---|---|---|---|---|
| *I.* | | | | | | | | |
| 1. | 5.13 miles | 9 miles. | 6.10 miles | 7 miles. | 40.50 m. | 15 miles. | 10.20 m. | 8.81 m. |
| 2. | 5.13 miles | 13 miles. | .......... | 3,000 ft. | 41 miles. | 7½ miles. | 4.63 m. | 8.81 m. |
| 3. | 64 lbs. | 56 lbs. | 62to95 lbs | 45to70 lbs | 45to64 lbs | 45&52 lbs. | 45 lbs. | 50 lbs. |
| *II.* | | | | | | | | |
| 1. | $340,637 | $252,859 | $89,217 | $50,577 | $1,461,308 | $191,955 | $179,924 | $200,920 |
| 2. | Bonds, $236,000 Mis., $10,119 | $4,864 | $6,486 | $15,322 | $34,990 | $5,231 | $41,967 | $123,349 |
| 3. | $586,757 | $257,704 | $95,704 | $65,900 | $1,496,294 | $197,180 | $221,89[illegible] | $324,269 |
| *III.* | | | | | | | | |
| 1. | 6,812,759 | 5,057,191 | 1,784,346 | 386,234 | 29,500,000 | 3,886,314 | 3,506,117 | 3,4[illegible]2,768 |
| 2. | 5 cents. | 5 cents. | 5 cents. | .......... | Adults, 5, 8, 10. Ch.3 & 4 | Adlts. 5 Ch.3 cts | Thro.2[illegible] Way, propor. | 8 cents. |
| 3. | 57 min. | 40 min. | 47 min. | 50 min. | .......... | .......... | 1 h. 46 m. | 45 min. |
| *IV.* | | | | | | | | |
| 1. | 444 | 400 | 190 | 1 horse, 9 dum. cars | 1,895 | 301 | .......... | 281 |
| 2. | 58 | 40 | 20 | 24 Pass. 2 Frt. | 412 | 68 | ........ | 58 cars. 14 sleighs |
| *V.* | | | | | | | | |
| 1. | $729,754 | $1,749,554 | $515,786 | $69,890 | $1,090,855 | $813,278 (items 1–4) | $698,800 (items 1–4) | $216,814 |
| 2. | $171,510 | $28,523 | $443,122 | $48,164 | $600,543 | | | $160,464 |
| 3. | $93,959 | Extens'n. $23,595 | $22,600 | $127,993 | $733,401 (items 3–4) | | | $71,216 |
| 4. | $59,455 | .......... | $17,600 | .......... | | | | $80,488 |

| | Second Avenue Railroad. | Third Avenue Railroad. | Sixth Avenue Railroad. | B'way and 7th Av. R. R. | Eighth Avenue Railroad. | Dry Dock and East B'way Railroad. |
|---|---|---|---|---|---|---|
| ***VI. Operating Expenses.*** | | | | | | |
| 1. Repairs to road-bed rlw., incl. iron, rep. of bldings, fixtures, | $22,497 | $131,975 Rl. Est. $220,281 | $110,255 | $41,469 | $47,390 | $49,691 |
| 2. Taxes on real est.. | $3,900 | $25,128 | $22,781 | $14,875 | $22,852 | $8,839 |
| 3. Superintend., office exp., clerks, etc.... | $22,209 | $44,675 | $12,476 | $18,850 | $20,261 | $22,359 |
| 4. Conductors & driv. | $146,089 | $341,116 | $170,136 | $194,052 | $162,111 | $239,99 |
| 5. Watchmen, starters, switchmen, and roadmen........... | $29,160 | $224,93 | $49,32 | $18,875 | $15,274 | |
| 6. Repairs of cars.... | $23,608 | $49,633 | $23,616 | Snow— $8,157 $30.516 | $16,524 | Engine, $1,50 Cars, $45,795 |
| 7. Repairs of harness, | $3,764 | $8,885 | $5,749 | $4,989 | $6,709 | $4.994 |
| 8. Horseshoeing..... | $24,500 | $60,768 | $39,727 | $28,020 | $29,096 | $22,674 |
| 9. Horses or mules... | $28,302 | $131,628 | $106,337 | $48,600 | $63,200 | $49,114 |
| 10. Stable expenses.. | $59,362 | $8,493 | $71,539 | $63,501 | $57,376 | $3,202 |
| 11. Feed, hay, etc.... | $119,820 | $237,144 | $150,780 | $154,404 | $137,298 | $138,61 |
| 12. Fuel, gas, & lights | $6,149 | $18,747 | $6,732 | $8,577 | $7,080 | $5,55 |
| 13. Oil and waste.... | $501 | $2,750 | $318 | $1,311 | $755 | $ 5 |
| 14. Water tax, ins'ce. | $915 $3,438 | $4,762 | $8,258 | $4,85 | $5,220 | $1,09 $7,30 |
| 15. Law expenses.... | $2,476 | $22,746 | $2,150 | $4,954 | $6,997 | $2,14 |
| 16. Damages to persons and property. | $5,118 | $2,294 | $1,768 | $6,264 | $2,220 | $ ,41 |
| 17. Rents............ | $1,899 | $19,150 | $4,000 | ........... | $9,927 | $ ,81 |
| 18. Car licenses...... | .......... | .......... | $3,500 | ........... | $5,250 | ........... |
| 19. Advert. and print. | .......... | $4,751 | $328 | $67 | $375 | $17 |
| 20. Taxes on divid'ds. | .......... | .......... | .......... | $487 | .......... | ...... ... |
| 21. Contingencies.... | $7,358 | $1,600 | $4,219 | $3,334 | $2.559 | Snow— $10,21 $3,02 |
| 22. Total operating expenses........... | $511,073 | $1,591,516 | $794,484 | $655,709 | $648,476 | $334,99 |

| | 42d and Grand St. Ferry Railroad. | Bleecker St. and Fulton F. Railroad. | Ninth Avenue Railroad. | Brooklyn Bath and Coney Is. Railroad. | Brooklyn City Railroad. | Brooklyn City and Newtown Railroad. | Coney Is. and Brooklyn Railroad. | Buffalo Street Railroad Company |
|---|---|---|---|---|---|---|---|---|
| *VI.* | | | | | | | | |
| 1. | $9,019 | $4,121 | $6,950 | $15,101 | $61,879 | $4,648 (1–2) | $15,141 | $21,945 |
| 2. | $14,770 | .......... | $4,657 | $382 | $49,433 | | $1,346 | $457 |
| 3. | $15,511 | $7,427 | $3,762 | $4,486 | $38,869 | $5,250 | $3,000<br>$2,901 | $10,706 |
| 4. | $67,359 | $49,549 | $28,109 | $3,065 | $387,120 | $62,727 (4–5) | $54,280 (4–5) | $47,959 |
| 5. | $20,116 | $7,053 | $3,691 | $3,427 | $69,383 | | | .......... |
| 6. | $17,419 | $20,119 | $3,320 | $2,939 | $53,816 | $7,852 (6–8) | $12,790 | $13,784 |
| 7. | $2,302 | $2,551 | $793 | $365 (7–11) | $11,306 | | $1,560 | $2,720 |
| 8. | $10,656 | $10,991 | $4,182 | | $50,600 | | $5,683 | $7,684 |
| 9. | $20,300 | $44,471 | $1,306 | | $45,874 | $9,448 | $10,323 | $18,935 |
| 10. | $30,622 | $22,415 | $8,874 | | $104,027 | .......... | $13,784 | $27,836 |
| 11. | $50,097 | $35,838 | $22,752 | | $245,313 | $36,033 | $34,338 | $23,881 |
| 12. | $2,338 | $1,541 | $739 | $2,753 | $8,360 | $935 | $1,158 | $1,879 |
| 13. | $46 | $484 | $104 | $661 | $627 | $103 | $273 | $261 |
| 14. | $1,583 | $1,932 | $1,352 | $1,139 | $9,365 | $1,526 | $3,11 | $2,794 |
| 15. | $4,250 | $390 | $330 | .......... | $3,000 | $50 | $654 | $1,456 |
| 16. | $1,588 | $632 | $448 | .......... | $16,781 | $318 | $600 | $528 |
| 17. | $1,500 | $6,022 | .......... | $150 | $2,328 | $401 | $237 | .......... |
| 18. | .......... | .......... | $400 | .......... | .......... | .......... | .......... | .......... |
| 19. | $[illegible] | $9[illegible] | $55 | $34 | $2,389 | .......... | $624 | $3,019 |
| 20. | .......... | .......... | .......... | .......... | .......... | .......... | .......... | .......... |
| 21. | $24,410 | $4,452 | $108 | $831 | $29,620 | $1,769 | $4,034 | $9,010 |
| 22. | $293,709 | $220,007 | $91,942 | $35,340 | $1,190,098 | $131,064 | $155,863 | $194,862 |

TABLE II.—*Giving particulars of the horse railways named below, compiled from the report of the Board of Railway Commissioners of the State of Massachusetts for* 1873.

| | Highland St. R'lwy Boston. | Lynn and Boston Railway. | Lowell Horse Railway. | Merimac Valley Hse. R'lwy Co. | Metropolitan Street R.'lway Co. |
|---|---|---|---|---|---|
| ***I. Description of Road*** | | | | | |
| 1. Length, miles | 5.42 | 11.75 | 3.815 | 5 | 43.608 |
| 2. Length of double track. | 1.985 | .13 | ...... | ...... | 8.734 |
| 3. Length of single track operated in one direction, miles | 1.17 | ...... | ...... | ...... | 5.097 |
| 4. Length of single track in both directions, miles. | .28 | 11.61 | 3.815 | 5 | 21.043 |
| 5. Length of switches, sidings, etc., miles | .25 | .87 | .216 | 1600 feet | 3.981 |
| 6. Length measured as single track, miles | 5.67 | 12.61 | 4.031 | 5.303 | 47.589 |
| 7. Weight of rail per yard, pounds | .48 | ¼ = 45 ¾ = 25 | 28½ to 33 | 1600 ft.=19 10400 ft.=30 | 30 to 55½ |
| ***II. Equipment.*** | | | | | |
| 1. No. of horses or mules | 252 | 239 | 50 | 55 | 1269 |
| 2. Number of cars | 36 | 35 | 12 | 15 | 201 |
| ***III. Cost of Construction.*** | | | | | |
| 1. Grading and Paving | $124,591 | ...... | $14,006 | $2,000 | ........ |
| 2. Track, including timber, rails, and laying | ...... | ...... | 37,211 | 35,784 | ........ |
| 3. Engineering and other exp. during construction, | 14,950 | ...... | 238 | ........ | ........ |
| 4. Total cost of construct'n | 139,541 | $181,960 | 51,455 | 37,784 | *1,100,437 |
| 5. Average per mile of sin. track, not incl. sidings | 25,745 | 15,485 | 13,487 | 7,556 | 25,235 |
| ***IV. Cost of Equipment*** | | | | | |
| 1. Horses or mules | 38,870 | 32,265 | 8,002 | 9,327 | 165,819 |
| 2. Cars | 41,605 | 30,300 | 12,002 | 16,618 | 189,956 |
| 3. Other vehicles and articles of equipment | 16,636 | 16,698 | 4,757 | 5,165 | 112,429 |

| | Highland St. R'lwy, Boston. | Lynn and Boston Railway. | Lowell Horse Railway. | Merrimac Valley Hse. R'lwy Co. | Metropolitan Street R'lwy Co. |
|---|---|---|---|---|---|
| ***V. Operating Expenses.*** | | | | | |
| 1. Reps. of rd.-bed & tracks | ...... | $15,767 | $1,845 | $2,507 | $17,915 |
| 2. Reps. of cars and other vehicles, harness, horse-shoeing.................. | $6,175 | 19,164 | 3,751 | 5,419 | 39,732 |
| 3. Repairs of buildings.... | ...... | 1,588 | 70 | 338 | 6,940 |
| 4. Keeping good the stock of horses ............... | 1,950 | 9,439 | ...... | 439 | 29,464 |
| 5. Wages of all employees. excepting Pres., Treas., Sup't. and clerks......... | 60,893 | 55,716 | 11,677 | 11,853 | 431,225 |
| 6. Provender .... ......... | 25,896 | 31,397 | 10,066 | 8,485 | 158,338 |
| 7. Taxes and insurance.... | 2,312 | 1,729 | 865 | 900 | 24,900 |
| 8. Damages for injuries to person and property ..... | 32 | 11,562 | 782 | ........ | 31,970 |
| 9. Rents & tolls paid other companies ............... | ...... | 18,402 | ...... | ........ | 3,019 |
| 10. Salaries, office exp., etc. | 5,960 | 12,013 | 3,068 | 4,245 | 117,673 |
| 11. Interest ................ | ...... | 5,051 | ...... | ........ | ........ |
| 12. Total exp. of operating. | 103,920 | 181,823 | 32,127 | 34,188 | 891,220 |
| ***VI. Revenue.*** | | | | | |
| 1. Received from pass'gers. | 127,399 | 162,713 | 33,555 | 34,002 | 945,585 |
| 2. Rec. from sale of manure | 609 | 1,301 | 638 | 520 | 2,219 |
| 3. Inc. from other sources. | ...... | 255 | 531 | 322 | 26,048 |
| 4. Total income........... | 128,008 | 164,269 | 34,724 | 34,845 | 982,853 |
| 5. Pr. ct. of exp. to income, | 92.5 | 110.68 | 92.5 | 98. | 90.676 |
| ***VII. Operating Particulars.*** | | | | | |
| 1. No. miles run by cars... | 397,432 | 447,068 | 122,953 | 176,280 | 2,470,214 |
| 2. Aver. cost per mile, cts. | 26.12 | 40.67 | 26.1 | 19.4 | 36,078 |
| 3. No. of pass'gers carried. | 2,511,180 | 2,150,652 | 592,716 | 453,673 | 18,211,026 |
| 4. Rate of speed, incl.stops. Miles per hour........... | 6 | 6 | 5 | 5 | 5 to 6 |
| 5. No. of persons regularly employed by Company .. | 149 | 102 | 25 | 24 | 640 |
| 6. Rate of fare—cents.. | 6ct. ticket 5 | 4 to 25 | 4, 5, and 6 | 3 | 5 to 15 |

* This is the total cost to the Company of road built and purchased; the cash cost is estimated to be $1,046,473.79.

The Brooklyn City Railroad Company operates 41 miles of double track, radiating from the Fulton and Hamilton Avenue ferries through the city to its suburbs in various directions. The returns for the year 1875, show that the company owns 1,950 horses. Each working horse travels, upon an average, 16⅓ miles daily, which somewhat exceeds the average daily distance traveled on the New York city lines. The monthly feed bill amounts to about $19,000.

During the year 1875 the company lost 334 horses, and the yearly depreciation in the value of the live-stock was about 28 per cent., the average annual depreciation for a term of years being about 25 per cent. The number of men employed is 1,000.

The company carried during the year 29,000,000 passengers, at an average cost per passenger of $4\frac{21}{100}$ cents, or an aggregate cost of $1,221,592.

The gross receipts from passengers were $1,446,537.

Mr. Sullivan, the President, states that among the various pavements in use between the rails, upon the ten roads operated by his company, the one which is the easiest upon the horses, and therefore the best, is formed of small cobble stones, laid with only a slight fall from the centre to the sides; and that the horses should always be shod with flat shoes, rather broad at the heel and without calks, and without cutting away any of the frog, so that a portion of the weight shall come upon the frog, whenever the animal treads upon an even surface. It is a rare occurrence for any of the horses to become disabled by contracted feet.

THE END.

# INDEX.

## A.



## B.



## C.

## S.



## T.



## U.



## V.

www.ingramcontent.com/pod-product-compliance
Lightning Source LLC
LaVergne TN
LVHW010248110826
845151LV00004B/1420

* 9 7 8 1 4 2 5 5 2 3 1 6 9 *